Proceedings in Life Sciences

*VIII th Congress of the International Primatological Society,
Florence, 7–12 July, 1980*

Selected Papers, Part A: Primate Evolutionary Biology
© by Springer-Verlag Berlin Heidelberg 1981

Selected Papers, Part B: Primate Behavior and Sociobiology
© by Springer-Verlag Berlin Heidelberg 1981

Main Lectures: Advanced Views in Primate Biology
© by Springer-Verlag Berlin Heidelberg 1982

Primate Evolutionary Biology

Selected Papers (Part A) of the VIIIth Congress
of the International Primatological Society,
Florence, 7–12 July, 1980

Edited by
A. B. Chiarelli and R. S. Corruccini

With 73 Figures

Springer-Verlag
Berlin Heidelberg New York 1981

Professor A. B. CHIARELLI
Istituto di Antropologia dell'Universita
Via del Proconsolo, 12
50122 Florence/Italy

Dr. R. S. CORRUCCINI
Istituto di Antropologia dell'Universita
Via del Proconsolo, 12
50122 Florence/Italy
and
Department of Anthropology
Southern Illinois University
Carbondale, IL 62901/USA

For explanation of the cover motive see legend to Fig. 2, p. 3.

ISBN 3-540-11023-2 Springer-Verlag Berlin Heidelberg New York
ISBN 0-387-11023-2 Springer-Verlag New York Heidelberg Berlin

Library of Congress Cataloging in Publication Data. Main entry under title: Primate evolutionary biology. (Proceedings in life sciences) Bibliography: p. Includes index. 1. Primates-Evolution-Congresses. 2. Primates-Morphology-Congresses. I. Chiarelli, A. B. II. Corruccini, Robert S. III. International Primatological Society. IV. Series. QL737.P9P6725. 599.8'0438. 81-14578. AACR2.

Offsetprinting and bookbinding: Brühlsche Universitätsdruckerei, Giessen
2131/3130-543210

Preface

The VIIIth International Congress of the International Primatological Society was held from 7 through 11 July 1980 in Florence Italy, under the auspices of the host institution, the Istituto di Antropologia of the University of Florence. More than 300 papers and abstracts were presented either at the main Congress or in 14 pre-Congress symposia the week earlier (so scheduled to avoid conflict with either the main invited lectures or the contributed paper sessions).

This volume consists of the contributed papers on primate evolutionary biology, primarily functional morphology, evolution, and paleontology. This is a coherent (though broad) and important subfield of primatology. We have thus focused the subject, in agreement with the publishers, to help ensure a successful and useful volume, befitting these very current contributions from the biennal meeting of the International Primatological Society.

Furthermore, we have compiled this volume in a fairly unprecedented manner for congress proceedings. In view of space and budget limitations, and the need to guarantee a high-quality book with appeal for specialists, we subjected all manuscript to a four-stage internal review process and selected only the best 13 of 31. This rejection rate of 58% compares with the more discriminating reviewed scholarly journals. Too often primatological or anthropological proceedings have been heterogeneous, large, unselective volumes that, at least recently, have repeatedly lost money for the publishing house (in sometimes spectacular amounts). Thus we offer a reasonably compact (therefore affordable) offering of the best recent research in this book; truly the "best of" the VIIIth IPC.

We have attempted a logical succession of papers in evolutionary morphology from lower Primates to the Apes and Man, although in this case the overlapping of functional and evolutionary themes made division by subject area impossible.

The editors jointly share in the responsibility for this volume, thus there is no junior editor. Our names appear at the front in alphabetical order.

We thank all for participating.

October, 1981

A.B. CHIARELLI
R.S. CORRUCCINI

Contents

The Homologies of the Lorisoid Internal Carotid Artery System
H. Butler (With 3 Figures).   1

Comparison of Eocene Nonadapids and *Tarsius*
P. Schmid (With 4 Figures)   6

Clinal Size Variation in *Archaeolemur* spp. on Madagascar
L.R. Godfrey and A.J. Petto (With 6 Figures)   14

The Anatomy of Growth and Its Relation to Locomotor Capacity
in *Macaca*
T.I. Grand (With 7 Figures).   35

Morphological and Ecological Characters in Sympatric Populations
of *Macaca* in the Dawna Range
A.A. Eudey (With 2 Figures)   44

Specialization of Primate Foot Reflected in Quantitative Analysis
of Arthrodial Joints of Anterior Tarsals
D.K. Messmann (With 11 Figures).   51

Morphology of Some of the Lower Limb Muscles in Primates
A. Prejzner-Morawska and M. Urbanowicz (With 9 Figures)   60

Morpho-Functional Analysis of the Articular Surfaces of the
Knee-Joint in Primates
C. Tardieu (With 7 Figures).   68

Outlines of the Distal Humerus in Hominoid Primates:
Application to Some Plio-Pleistocene Hominids
B. Senut (With 9 Figures)   81

Structural-Functional Relationships Between Masticatory Biomechanics, Skeletal Biology and Craniofacial Development in Primates
O.J. Oyen and D.H. Enlow (With 2 Figures)   93

Comparison of Morphological Factors in the Cranial Variation
of the Great Apes and Man
B. Jacobshagen (With 9 Figures).   98

Enamel Prism Patterns of European Hominoids — and Their
Phylogenetical Aspects
N.I. Xirotiris and W. Henke (With 4 Figures).   109

The Structural Organization of the Cortex of the Motor Speech
Areas of the Human Brain and Homologs on the Ape's Brain
M.S. Vojno.   117

Contributors

You will find the addresses at the beginning of the respective contributions

Butler, H. 1
Enlow, D.H. 93
Eudey, A.A. 44
Godfrey, L.R. 14
Grand, T.I. 35
Henke, W. 109
Jacobshagen, B. 98
Messmann, D.K. 51
Oyen, O.J. 93

Petto, A.J. 14
Prejzner-Morawska, A. 60
Schmid, P. 6
Senut, B. 81
Tardieu, C. 68
Urbanowicz, M. 60
Vojno, M.S. 117
Xirotiris, N.I. 109

The Homologies of the Lorisoid Internal Carotid Artery System

H. Butler [1]

In most primates the brain is supplied by two arterial systems, the vertebral-basilar and the internal carotid. The internal carotid or promontory artery enters the bulla and traverses the middle ear cavity (Fig. 1A). The pattern of the carotid artery system of the lorisoids is very different. The promontory artery is very small, reduced to a fibrous cord or absent (Packer and Conroy 1980). It is replaced by an artery arising at or near the bifurcation of the common carotid artery, which runs forward medial to the bulla (Fig. 1B). It enters the cranial cavity via the foramen lacerum to reach the middle cranial fossa, where it ends in the circulus arteriosus. This apparent internal carotid artery forms an extracranial rete mirabile, which appears to be best developed in the galagids (Ask-Upmark 1953, Adams 1957, Bugge 1974). The possible homologies of this artery were considered by Cartmill (1975) to be:

1. The internal carotid artery of Man, i.e., the promontory artery.
2. The medial entocarotid.
3. The ascending pharyngeal artery of Man.
4. A neomorph, not homologous with any vessel found in other primates or in primitive mammals.

Kanagasuntheram and Krishnamurti (1965) examined the development of the internal carotid artery system in the lesser galago *(Galago senegalensis senegalensis)* and concluded that the artery which replaced the degenerated segment of the internal carotid artery was probably the ascending pharyngeal artery.

The following account of the development of the internal carotid artery system of the lesser galago *(Galago senegalensis senegalensis)* is based on the examination of eight embryos and fetuses ranging from 8.5 to 26.0 mm C.R. length.

In this account, the first-appearing "internal" carotid artery, which runs through the middle ear and enters the middle cranial fossa to join the circulus arteriosus, will be referred to as the promontory artery. The artery which eventually replaces most of it will be referred to as the artery to the rete.

The promontory artery of the 8.5 mm C.R. length embryo runs forward in the roof of the pharynx between the otic capsule medially and the middle ear cavity laterally. It is widely patent and there is no sign of the rete. It gives off a large stapedial artery which passes through an opening in the mesenchymal anlage of the stapes.

1 Department of Anatomy, University of Saskatchewan, Saskatoon, Saskatchewan, Canada S7N OWO

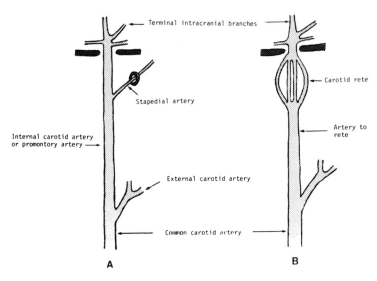

Fig. 1. A The internal carotid or promontory artery found in most primates. **B** The "internal carotid" artery of the lorisoids

The promontory artery then enters the cranial cavity lateral to the hypophysis and terminates by dividing into the middle cerebral, anterior cerebral, and two ophthalmic arteries. It is connected to the vertebral-basilar arterial system by the posterior communicating artery.

The rete and the artery to the rete are first seen in the 11.0 mm embryo and are fully developed in the 14.0 mm embryo. The artery to the rete arises from the wide-bored commencement of the promontory artery just beyond the bifurcation of the common carotid artery. It runs dorso-medially between the otic capsule and the roof of the pharynx to join the caudal end of the rete. The rete lies on the ventral aspect of the otic capsule and forms a sponge-like collection of arterial spaces mixed with wide venous spaces. After the origin of the artery to the rete the promontory artery is considerably narrower and runs cranialward between the lateral surface of the otic capsule and the middle-ear cavity. The promontory artery is accompanied by the internal carotid nerve which arises from the lateral aspect of the superior cervical ganglion. During this part of its course, the promontory artery gives off a small stapedial artery and then joins the cranial end of the rete. The wide intracranial part of the promontory artery arises from the cranial end of the rete and enters the cranial cavity through the foramen lacerum. Inside the cranial cavity it terminates by dividing into the middle cerebral, anterior cerebral, and two ophthalmic arteries and connects with the vertebral-basilar arterial system via the posterior communicating artery. Figure 3 shows the relationship of the various components of the "internal carotid" arterial system to the otic capsule, middle-ear cavity, pharynx, and chondrocranium.

The 22.5 mm C.R. length fetus shows the adult arterial pattern. The artery to the rete runs medially and cranially to join the caudal end of the rete. It supplies the

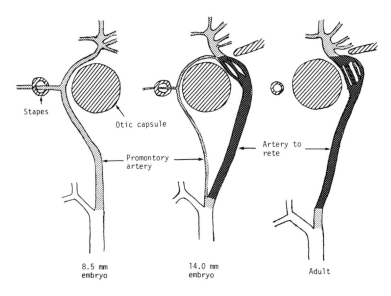

Fig. 2. Diagrammatic summary of the development of the "internal carotid" arterial system of the lesser galago

pharyngeal muscles and is crossed laterally by the stylopharyngeus muscle, and the glossopharyngeal and hypoglossal nerves. The cranial end of the rete lies in the gap between the anterior end of the otic capsule and the posterior margin of the ala temporalis, i.e., the foramen lacerum. The greater petrosal (Vidian) nerve runs across this gap close to the cranial end of the rete. The rete opens into the wide intracranial part of the promontory artery which enters the cranial cavity via the foramen lacerum and divides into its terminal branches.

The development of the "internal carotid" arterial system of the lesser galago is summarized diagrammatically in Fig. 2. It commences as a typical promontory artery running through the middle ear and crossing the lateral aspect of the otic capsule where it gives off a stapedial artery. It then enters the cranial cavity to terminate by dividing into the middle cerebral, anterior cerebral, and dorsal and ventral ophthalmic arteries. It is connected to the vertebral-basilar arterial system by the posterior communicating artery.

The artery to the rete arises from the commencement of the promontory artery and runs cranially and medially, passing the medial aspect of the otic capsule to join the caudal end of the rete. As the reduced promontory artery enters the foramen lacerum, it joins the cranial end of the rete. Its wide intracranial terminal part forms the efferent artery of the rete.

The proximal part of the promontory artery, between the origin of the artery to the rete and its connection with the cranial end of the rete, eventually disappears as does its stapedial branch. Thus the vessel often identified as the internal carotid artery in the adult is a compound vessel formed from the commencement of the promontory artery, the artery to the rete, the rete and the terminal intracranial part of the promontory artery.

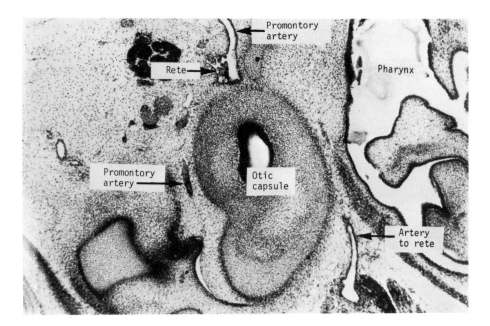

Fig. 3. *Galago senegalensis,* 14.0 mm C.R. length embryo. This section shows the relationship of the component parts of the lorisoid "internal carotid" arterial system to the otic capsule

Thus it is clear that the "internal carotid" artery of the lesser galago is only partially derived from the promontory artery, viz. its commencement and its terminal intracranial part. The part which disappears is replaced by a new artery which grew out of it to join the rete which, in turn, joins the terminal intracranial part. Presley (1979) indicated that two arteries were never present in the various embryos that he examined and that a single dorsal aorta-derived internal carotid artery develops in an intermediate position and, as a result of differential growth of the basicranial elements, comes to lie medial or lateral to the bulla. The present observations substantiate Presley's (1979) view that the internal carotid artery system of mammals, including the lesser galago, commences as a single dorsal aorta. In the lesser galago, however, there is a stage when two "internal carotid" arteries are present, one medial to the bulla and one lateral to it (Fig. 3). This does not rule out the possibility that the original single dorsal aorta-derived internal carotid artery may shift, medially or laterally, as a result of differential basicranial growth in other mammals.

These observations fully confirm and expand the findings and conclusions of Kanagasuntheram and Krishnamurti (1965). They also corroborate the use of the internal carotid nerve as a marker for the extracranial part of the promontory artery (Packer and Conroy 1980).

The relationships of the artery to the rete indicate that it is the homolog of the human ascending pharyngeal artery. Cartmill (1975) notes the following features of the human ascending pharyngeal artery:

1. It supplies pharyngeal muscles and the soft palate.
2. An accessory meningeal branch passes through the foramen lacerum close to the deep petrosal (Vidian) nerve where the latter crosses the foramen lacerum on its way to the pterygopalatine (Vidian) canal. This artery may anastomose with branches of the internal carotid artery that supply the trigeminal ganglion.
3. It arises from the internal or external carotid arteries or the carotid bifurcation.
4. It is crossed laterally by the stylopharyngeus muscle and the glossopharyngeal and hypoglossal nerves.

The artery to the rete supplies pharyngeal muscles. The cranial end of the rete lies in the foramen lacerum and is crossed by the deep petrosal (Vidian) nerve. The cranial end of the rete joins its terminal intracranial part in the foramen lacerum, i.e., between the otic capsule and the alisphenoid. The medial entocarotid artery, however, enters the cranial cavity between the petrous temporal and occipital bones. The coincidence of the relationships of the artery to the rete with the human ascending pharyngeal artery would appear to rule out the possibility that it is a neomorph.

References

Adams WE (1957) The extracranial carotid rete and carotid fork in *Nycticebus coucang.* Ann Zool 3:21–38

Ask-Upmark E (1953) On the entrance of the carotid artery into the cranial cavity in *Stenops gracilis* and *Otolicnus crassicaudatus.* Acta Anat 19:101–103

Bugge J (1974) The cephalic arterial system in insectivores, primates, rodents and lagomorphs with special reference to the systematic classification. Acta Anat 87 (Suppl 62):1–60

Cartmill M (1975) Strepsirhine basicranial structures and the affinities of the cheirogaleidae. In: Luckett WP, Szalay FS (eds) Phylogeny of the primates. Plenum Press, New York London, pp 313–353

Kanagasuntheram R, Krishnamurti A (1965) Observations on the carotid rete in the lesser bush baby *(Galago senegalensis senegalensis).* J Anat 99 (4):861–875

Packer DJ, Conry GC (1980) Carotid artery homologies in strepsirhine primates. Am J Phys Anthropol 52:265

Presley R (1979) The primitive course of the internal carotid artery in mammals. Acta Anat 103: 238–244

Comparison of Eocene Nonadapids and Tarsius

P. Schmid [1]

Views on the relationships of the Eocene non-adapid primates have flucutated moderately through time (Szalay 1975a). The creation of family concepts for the Omomyidae Trouessart, 1879 (including Anaptomorphidae Cope, 1883 and Microchoeridae Lydekker, 1887) signifies that these groups were previously recognized as distinct from the Adapidae Trouessart, 1879.

Wortman (1903) asserted the closer ties of *Tarsius* and the nonadapid primates, establishing the "Paleopithecini" (Tarsiidae and Anaptomorphidae). Simpson (1940) came to doubt the clear-cut differentiation of tarsioid-lemuroid traits in Eocene primates. Emphasizing the taxonomic characters of the ear region, Hürzeler (1948) suggested a close relationship of *Necrolemur* and relatives with the Lemuriformes, rather than with the living *Tarsius*. In 1961 Simons took the step of including the microchoerines in the Tarsiidae. A few years later, Robinson (1968) made the unusual proposition of aligning the Omomyidae with the Lorisiformes. In an extensive discussion of the phylogeny of the Tarsiiformes, Szalay (1975a) stated "The primitive omomyid character states of many known features are derived compared to the more primitive conditions of the adapid morphotype".

The main question this paper is concerned with is the relationship of Eocene nonadapids to the extant *Tarsius*. The fossil forms are only represented by the skulls of *Tetonius, Rooneyia*, and *Necrolemur*. Our discussion is based on the material of the latter genus, which we were able to study in Paris (Muséum National d'Histoire Naturelle) and London (British Museum of Natural History).

The primate subordinal division and the allocation of *Tarsius* was discussed at the congress of the IPS at Cambridge (see Chivers and Joysey 1978). The strepsirhine-haplorhine model was accepted by a large number of scientists. Besides fetal membranes, placental characters (Luckett 1975), and immunological data (Goodman et al. 1978), there are some osteological features that serve to determine which fossil form belongs to the haplorhine clade. The separation of the Strepsirhini and Haplorhini should be recognizable in regarding the Eocene fossils, where we should be able to distinguish between Adapiformes and Tarsiiformes.

---

1 Anthropologisches Institut der Universität Zürich, Künstlergasse 15, 8001 Zürich, Switzerland

Character Analysis

In addition to an amount of nonosteological data, the haplorhine morphotype is clearly separated from the strepsirhine clade by several derived characters (Table 1). In weighing these characters, which were proposed to separate the two groups (Luckett and Szalay 1978), we try to analyze the cladistic relationship of the fossil so-called tarsiiforms from the Upper Eocene.

Table 1. List of derived character states in osteological features of the Haplorhini (slightly modified after Luckett and Szalay 1978). + = derived condition, – = primitive condition

	Strepsirhini	Necrolemur	Tarsius	Simii
1. No gap between the medial upper incisors	–	–	+	+
2. Foramen caroticum posteromedial to the bulla	+ –	+	–	+
3. Pneumatization of the bulla	+ –	+	+	+
4. Promontory artery well developed	–	–	+	+
5. Stapedial artery reduced	+ –	–	+	+
6. Fenestra rotunda ventrally exposed	–	–	+	+
7. Extrabullar ectotympanic	+ –	+	+	+
8. Separation of the orbital fossa and the temporal fossa	–	–	+	+
9. Olfactory process above the interorbital septum	–	+	+	+

No Gap Between the Medial Upper Incisors

The tooth-combed prosimians possess a naked rhinarium. The ventral part of the rhinarium (= philtrum) passes through a gap between the medial upper incisors. This character state is widely distributed and considered as primitive or plesiomorph. On the contrary, there is a derived or apomorph character state of a movable hairy upper lip in *Tarsius* and the other Haplorhini (Pocock 1918), They show a crowded incisor region in the upper jaw (Maier 1979). In all fossil Eocene primates, where this part is preserved, we can find a considerable gap between the upper front teeth. So we may suppose a naked rhinarium in all these forms including *Necrolemur* and *Microchoerus*.

Foramen Caroticum Posteromedial to the Bulla

Szalay (1972, 1975b) proposed a morphocline of this feature (Fig. 1). He regarded the carotid foramen on the posterolateral side of the inflated bulla, near the opening for the facial nerve, as the primitive character state in the primates. This is found in *Plesiadapis* and *Phenacolemur,* in the adapids, and in the recent lemurids. A shift of this carotid entry in *Necrolemur* to the medial side of the bulla was interpreted as an intermediate state, where the opening in the center of the bulla, as shown by *Tarsius,* was considered as a derived feature.

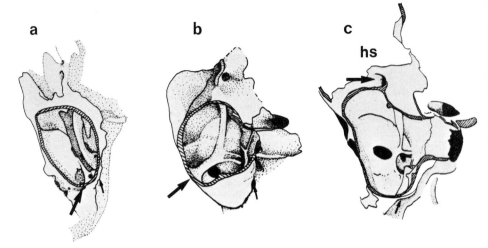

Fig. 1 a–c. Proposed morphocline of the carotid entry by Szalay (1972, 1975b). a *Lepilemur* (primitive), b *Necrolemur* (intermediate), c *Tarsius* (derived). We consider the widely distributed condition in b as primitive, whereas a is the derived state of the lemurids and c the derived condition of *Tarsius*. *Big arrow,* Foramen caroticum; *small arrow,* opening for the facial nerve; *asterisk,* bony septum which shields the Fenestra rotunda; *hs,* hypotympanic sinus. (The bullas are not equally enlarged and in c only the posterior part of the bulla is represented)

In comparison with other primates, we found a medial entry of the carotid artery in the galagids, lorisids, callitrichids, cebids, and the catarrhines. Also a number of other mammals show this character state, so we consider this as a symplesiomorph feature. The outgroup comparison indicates that only the lemurs and some plesiadapoids have a posterolateral carotid foramen, which represents an apomorph or derived state. In adult *Tarsius* the Arteria promontorii never touches the cochlea. This, on the other hand, is an autapomorph character for this genus, which is not shared by *Necrolemur* nor by *Rooneyia*.

Pneumatization of the Bulla

The bullar portion of the petrosal is differently pneumatized in the various lineages of the primates. A morphocline can hardly be proposed. As an example, in the strepsirhines there can be seen two different manners of bullar pneumatization.

In the lemurs, the bulla is ventrally inflated. The petrosal plate does not fuse with the ectotympanic and grows below and lateral to the tympanic ring. With this growing tendency also the carotid foramen moves to a lateral position (Fig. 2B).

Not so in the lorisoids, where the bulla is pneumatized in another way. The petrosal plate fuses with the ectotympanic, which forms the lateral bullar wall (Fig. 2C). The carotid foramen remains in a primitive position near the cochlea. In addition, this vessel is largely reduced and replaced by an ascending Arteria pharyngea.

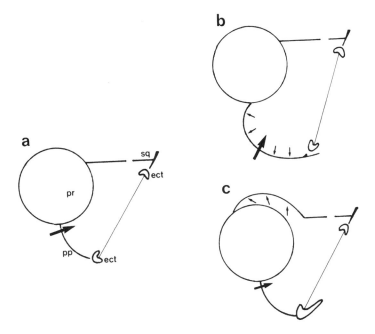

Fig. 2 a–c. Different bullar ontogeny in strepsirhine primates (after MacPhee 1977). **a** neonatal, **b** *Microcebus*, **c** *Galago*. *Small arrows* indicate the sites of pneumatic activity. *Bigger arrows* show the schematic position of the carotid entry. *Ect*, ectotympanic ring; *pp*, petrosal plate; *pr*, promontorim; *sq*, squamosum

Promontory Artery Well Developed

In the primate morphotype the internal carotid divides on the cochlea to the A. promontorii and the A. stapedia. These vessels are housed in bony tubes or leave impressions on the bony wall of the promontorium. After the bifurcation, the vessels have reduced diameters, which vary slightly. This character state is represented by *Lemur catta* for example. In *Phaner*, where the A. stapedia tends to reduce, the A. promontorii remains a weak vessel because the main blood supply is primitively furnished by the A. vertebralis (Saban 1975).

Only in the recent haplorhines the diameter of the internal carotid and the A. promontorii remains the same size, because the importance for the cranial irrigation is emphasized.

Necrolemur from the Eocene period shows the primitive primate condition with reduced calibres after the bifurcation of the internal carotid.

Stapedial Artery Reduced

The primitive character state of this feature is mentioned above and found in most of the Lemuridae and the Paleogene fossils. The derived feature is the reduction of one or both vessels of the intrabullar circulation. The blood supply is assured by the development of anastomoses (Bugge 1972).

The latter character state is found in the Cheirogaleidae and the lorisoids, where both branches of the internal carotid are reduced. The meningeal and ophthalmic areas are supplied by the ascending A. pharyngea (Conroy and Packer, in press).

The stapedial artery is reduced in *Tarsius* and in the simian families. The A. promontorii remains the main vessel and shows a constant diameter after the branching point. The stapedial artery can only be found in early ontogenetical stages, but obliterates very soon.

Necrolemur shows the primitive condition, where the main vessel divides into two channels with small diameter differences.

Fenestra Rotunda Ventrally Exposed

Where the primitive circulation pattern is found, the bony tube of the carotid passes near the Fenestra rotunda.. Sometimes the channel is enclosed in a bony septum. Where this bony rim of the petrosal is found, the Fenestra rotunda is ventrally covered. This is the case in all strepsirhines and also in plesiadapoids, omomyids, microchoerids and adapids. Whereas in haplorhines this structure is missing and the opening of the cochlea is exposed.

Extrabullar Ectotympanic

An extrabullar ectotympanic is widely distributed in fossil and extant primates. Conroy (1980) suggested that the ectotympanic tube or ossified annulus membrane found in Early Tertiary primates was not necessarily a homologous structure to the true ectotympanic tube seen in haplorhines, and thus could not be considered as a shared derived feature linking known Paleogene primates from Europe and North America to tarsioid and anthropoid ancestry.

Separation of the Orbital Fossa and Temporal Fossa

Primitively the strepsirhine character state is marked by a single bony bar or ring that defines but does not separate the orbital fossa from the temporal fossa.

In the Haplorhini, we can see the derived condition, where the postorbital opening is partially closed by bony protrusions of the malar, maxillary, frontal, and alisphenoid.

In *Necrolemur*, this feature is plesiomorph. The flanges of the postorbital bar are not more pronounced than in *Lepilemur* for example.

Olfactory Process Above the Interorbital Septum

In the Strepsirhini the olfactory bulb passes below the separation of the orbits, whereas in the Haplorhini the process runs above the bony separation (Fig. 3).

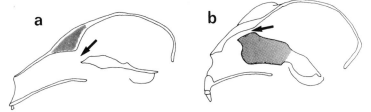

Fig. 3. Relationships of the orbital septa in **a** the strepsirhines, and **b** the haplorhines. Interorbital septum *stippled, arrows* indicate the direction of the olfactory process

This is the only osteological character which seems to be shared by the omomyids, the microchoerids, the tarsiids and the higher primates. An exact analysis of this region in fossils is not possible, because it involves the destruction of the material.

Conclusions

The foregoing character analysis of osteological features corroborates the hypothesis of close relationships within the recent haplorhine primates. The Haplorhini (including recent *Tarsius*, New and Old World monkeys, and apes) is a monophyletic taxon defined by its common possession of shared and derived homologous characters.

The hypothesis of an early (Paleocene) dichotomy of strepsirhine and haplorhine primates is based on a clear-cut distinction of the two lineages within the Eocene fossil record. Our analysis indicates that the Tarsiiformes (including Omomyidae, Microchoeridae, and Tarsiidae) do not form a sister group based on shared derived characters. Therefore we exclude the Eocene microchoerids from the infraorder. Figure 4 illustrates two possibilities of non-adapid relationships. Cladogram A is falsified by the present investigation, whereas diagram B is the most likely hypothesis of microchoerine (and probably omomyine) relationship.

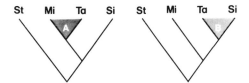

Fig. 4. A cladogram showing a close relationship of the Microchoeridae to the Tarsiidae which together form a "tarsiiform" clade. **B** most likely hypothesis of microchoerine relationship. *Mi,* Microchoeridae; *Si,* Simii; *St,* Strepsirhini; *Ta,* Tarsiidae

In most of the osteological characters analyzed, *Necrolemur* shows the primitive condition. No feature is shared by *Tarsius* and *Necrolemur* exclusively. Only the relationship between the olfactory bulb and the interorbital septum seems to represent a haplorhine apomorphism (Cave 1967, Cartmill 1972a,b) which is also shared by the fossil non-adapids. Such investigations about the paranasal sinus pattern in fossil forms are lacking, because they involve the destruction of the skull.

At present, it would be wise to consider *Tarsius* as a very specialized haplorhine primate, which has not much to do with the haplorhine morphotype.

It remains unclear, if the non-adapid Omomyidae (excl. *Rooneyia*) and Microchoeridae show evidence for being such a basal form. All we can conclude is that the Adapiformes are clearly separated from a radiation of Paleogene primates by the laterally shifted Foramen caroticum, a ringlike ectotympanic, and an extension of the tympanic cavity laterally to the ring.

Acknowledgments. I am grateful to Prof. F.S. Szalay for his critical comments (although he does not agree with my general conclusions). I would also like to acknowledge helpful discussion with and encouragement by Prof. W.P. Luckett and Prof. Dr. med J. Biegert. Thanks are due to Prof. J.-P. Lehmann (Muséum national d'Histoire Naturelle, Paris) and Dr. J. Hooker (British Museum of Natural History, London) for permission to study collections in their care. Financial support was provided by the A.H. Schultz-Foundation.

References

Bugge J (1972) The cephalic arterial system in the insectivores and the primates with special reference to the Macroscelidoidea and Tupaioidea and the insectivore-primate boundary. Z Anat Entwicklungsgesch 135:279–300

Cartmill M (1972a) Arboreal adaptations and the origin of the order primates. In: Tuttle (ed) The functional and evolutionary biology of primates. Aldine, Chicago

Cartmill M (1972b) Strepsirhine basicranial structures and the affinities of the cheirogaleidae. In: Luckett WP, Szalay FS (eds) Phylogeny of the primates. Plenum Press, New York London

Cave AJE (1967) Observations on the platyrrhine nasal fossa. Am J Phys Anthropol 26:277–288

Chivers DJ, Joysey KA (eds) (1978) Recent advances in primatology, vol III. Evolution. Academic Press, London New York

Conroy GC (1980) Ontogeny, auditory structures, and primate evolution. Am J Phys Anthropol 52:443–451

Conroy GC, Packer DJ (in press) The anatomy and phylogenetic significance of the carotid arteries and nerves in strepsirhine primates. Folia Primatol

Goodman M, Hewett-Emmett D, Beard JM (1978) Molecular evidence on the phylogenetic relationships of *Tarsius*. In: Chivers GC, Joysey KA (eds) Recent advances in primatology, vol III. Academic Press, London New York

Hürzeler J (1948) Zur Stammesgeschichte der Necrolemuriden. Schweiz Palaeontol Abh 66:1–46

Luckett WP (1975) Ontogeny of the fetal membranes and placenta – their bearing on primate phylogeny. In: Luckett WP, Szalay FS (eds) Phylogeny of the primates. Plenum Press, New York London

Luckett WP, Szalay FS (1978) Clades versus grades in primate phylogeny. In: Chivers DJ, Joysey KA (eds) Recent advances in primatology, vol III. Academic Press, London New York

MacPhee RDE (1977) Ontogeny of the ectotympanic-petrosal plate relationship in strepsirhine prosimians. Folia Primatol 27:245–283

Maier W (1979) *Macrocranion tupaiodon*, an adapisoricid (?) insectivore from the Eocene of "Grube Messel" (Western Germany). Palaeontol Z 53:38–62

Pocock RI (1918) On the external characters of the lemurs and of *Tarsius*. Proc Zool Soc London, pp 19–53

Robinson T (1968) The paleontology and geology of the Badwater Creek area, central Wyoming, part 4. Late Eocene primates from the Badwater, Wyoming, with a discussion of the material from Utah. Ann Carnegie Mus 39 (19):307–326

Saban R (1975) Structure of the ear region in living and subfossil lemurs. In: Tattersall I, Sussman RW (eds) Lemur biology. Plenum Press, New York London

Simons EL (1961) Notes on Eocene Tarsioids and a revision of some Necrolemurinae. Bull Br
    Mus Nat Hist Geol 5 (3):47–70
Simpson GG (1940) Studies on the earliest primates. Bull Am Mus Nat Hist 77:185–212
Szalay FS (1972) Cranial morphology of the Early Tertiary *Phenacolemur* and its bearing on
    primate phylogeny. Am J Phys Anthropol 36:59–76
Szalay FS (1975s) Phylogeny, adaptations, and dispersal of the Tarsiiform primates. In: Luckett
    WP, Szalay FS (eds) Phylogeny of the primates. Plenum Press, New York London
Szalay FS (1975b) Phylogeny of primate higher taxa – the basicranial evidence. In: Luckett WP,
    Szalay FS (eds) Phylogeny of the primates. Plenum Press, New York London
Wortman JL (1903) Studies of Eocene mammalia in the Marsh Collection, Peabody Museum,
    part 2. Primates. Am J Sci 15:163–176

Clinal Size Variation in Archaeolemur spp. on Madagascar

L.R. Godfrey and A.J. Petto [1]

The Problem of Scale

Researchers in primatology have historically viewed size as a confounding variable in the analysis of skeletal remains. A decade ago it was common knowledge that in order to discover what was interesting about morphological variation one had to somehow "get rid of" size variation ("correct" for variation in shape which depends upon size). However, in more recent years attention has increasingly focused on size as "adaptive strategy". The study of scale has influenced almost every aspect of primatological research, as, for example, research on postcranial morphology and positional behavior (Cartmill 1974, Rose 1974, Fleagle 1979), dental morphology and diet (Kay 1975, Pirie 1978), activity rhythms (Charles-Dominique 1975), the evolution of the brain (Stephan and Bauchot 1965, Jerison 1973, Passingham 1975, Pilbeam and Gould 1974, Gould 1975), gestation and reproductive behavior (Leuten-egger 1973, Sacher and Staffeldt 1974, Martin 1975), ranging behavior, social structure and ecology (Milton and May 1976, Clutton-Brock and Harvey 1977, Ripley 1979), and phylogenetic reconstruction (Martin 1979).

Several factors have contributed to this change in research orientation. Of foremost importance has been the development in the fields of physiology, biomechanics, and biomathematics of a science of "scaling", focusing on rules governing changes in shape associated with changes in scale (Bonner 1965, 1974, Schmidt-Nielsen 1970, 1972, Taylor et al 1972, McMahon 1975, Gould 1966, 1970, 1971, 1977, Alexander and Goldspink 1977, Pedley 1977). Organisms have a limited variety of ways of coping with weight-related stresses. Given the constraints of biological materials of which organisms are constructed, closely related or adaptively similar organisms often cope with changes in scale by maintaining some pattern of "functional" similarity while sacrificing others [2] . Such patterns are definable and often predictable within

---

1 Department of Anthropology, University of Massachusetts, Amherst, MA 01003, USA

2 It is mathematically impossible for isometry to be maintained in all parameters of three-dimensional organisms of different sizes. Even when the relationship between any two linear dimensions (or lengths and diameters) is isometric, the relationship between volumes and areas, or volumes and linear dimensions, will not be isometric. Two organisms cannot be identical in all proportions or ratios unless they are the same size. Thus, if structural or functional similarity is maintained in some aspect, it must be sacrificed in another. Furthermore, structural and functional similarity are not the same; if two organisms are geometrically similar, the larger

the rubric of physics, and a few basic "laws" of scaling tend to operate as major influences on morphology and physiology (cf. McMahon 1975). Changes in morphology and physiology impose constraints upon behavior and ecology. Therefore, morphology, physiology, behavior, and ecology are all better understood when problems of scaling are taken into account. This is the primary level of analysis of scale: discovering and predicting the *nature* of allometric (size-required) constraints upon form.

Yet beyond such descriptions lie important theoretical challenges, for if we are to understand not merely the scale requirements to which selection must respond, but also the *processes of change* which lead to the successful acquisition of these adaptations, we must study allometric variation in ontogenetic, intrageneric, and intergeneric (phylogenetic) series, and develop testable hypotheses or models of genetic control. This secondary level of analysis of scale addresses two distinct problems:

1. How do developmental processes limit or control phylogenetic change? How do regulatory processes work? What are the internal constraints upon evolutionary change in complex adaptations? (cf. Løvtrup 1974, Frazzetta 1975, Waddington 1975, Gould 1977).

2. How are these processes of change limited by population parameters such as gene flow, population structure, deme size, etc.? Are certain kinds of regulatory gene change only possible in small founder populations?

Such analysis bears directly on controversies over evolutionary processes. For example, the punctuationalist argument that certain kinds of changes are only possible in association with speciation (Eldredge and Gould 1972, Gould and Eldredge 1977, Stanley 1975, 1979) can be tested by a careful examination of variation at different levels of organization.

A third level of analysis of scale establishes *external* constraints on change in size: under what conditions would shifts in size be favored? This is the study of size as adaptive strategy. The critical question here is not constraints of size on shape per se, but constraints of the external environment on size and shape. Certain generalizations (such as the Bergmann and Allen rules) come to mind immediately, but examples of shifts in size which do not conform to the expectations of the Bergmann-Allen rules are ubiquitous. In fact, many often directly conflicting explanatory models for change in size have been posited (compare, for example, the discussions of scale in Kurtén 1953, Brody 1945, Hairston et al. 1970, Taylor et al 1970, Pianka 1970, 1972, Geist 1971, Wilson 1975, Bourlière 1975, Grubb 1972, Pirie 1978, Wassersug et al. 1979). It is certain that multicausal models will be required. Endler (1977) in analyzing clines, and Gould (1977) in analyzing ontogenetic and phylogenetic changes in form, have shown how very different *processes* can produce virtually indistinguishable *results*. It is clear that we need to do more than acknowledge the *existence* of changes in scale if we wish to understand them. We must address methodological problems in studying the ecological context of variation in size, and

---

(Footnote 2, continued)

will sacrifice some capacity to withstand the compressive and bending stresses imposed under its own body weight. If elastically similar, the larger will maintain such capacity, but sacrifice similarity in shape. Functional similarity is thus constrained by those very corollaries of maintaining any given constant relationship or similarity

focus attention on developing an adequate data base to test alternative processural models.

This paper focuses on this "third" level of analysis of change in scale, but for a special case: extinct organisms. Our objective is to demonstrate the potential of a series of mathematical tools (especially factor analysis, contour mapping, and trend surface analysis): first, to establish the existence of a size cline; second, to describe the nature of the size cline; and third, to explore associated changes in the environment.

Clinal Variation in Fossil Forms: the Case of Archaeolemur

Extinct organisms can provide information generally not available to biologists through other means. Not only do they provide a diachronic perspective essential to the understanding of evolutionary processes, but they form a valuable data base for studying processes of genetic differentiation in environments less disturbed by humans than those which exist today. But their fragmentary nature has generally discouraged analysis of intrageneric variation in form. Historically, with few exceptions, researchers in paleontology have tended to focus on *description* and *taxonomic assignment* rather than *spatial* variation and processes of evolutionary change. Even the question of function – reconstructing "faculty" (Bock and von Wahlert 1965, Morbeck et al 1979) or "totipotential" behavior (Prost 1980) – has been largely ignored until very recently.

Endler's (1977) study of clinal variation in extant organisms demonstrated that speciation can occur through a number of very different pathways. Rapid speciation via genetic processes *unrelated* to gradual adaptive allelic substitution is now widely recognized as possible (even without allopatric separation of demes) (cf. Carson 1973, Gottlieb 1975, White 1978, Gould 1980). Given gradual adaptive allelic substitution it was, practically speaking, impossible to *define* the point at which a new species emerges (either for cladogenesis or anagenesis). Any measure or morphological change could only yield an approximate index of genetic differentiation. However, even if the punctuationalists are *partly* right (if certain kinds of morophological change are commonplace within lineages, and other kinds of morphological change can occur in association with rapid speciation or genetic isolation), the following research question becomes nontrivial: Under what conditions of morphological divergence can we safely reject the "null" hypothesis that reproductive isolation has not occurred?

The point is this: for closely related organisms we cannot know whether speciation has occurred *unless* we know the *extent* to which a certain observed pattern of morphological variation *precludes* or *requires* obstruction of gene flow. The question which should be posed first is not how many species are represented by a set of fossils, but what is the pattern of morphological (spatial and temporal) variation exhibited by the set of fossils, and what are the possible genetic models which can account for it? This encompasses questions of population structure, gene flow, spatial distribution of demes, and underlying patterns of ontogenetic change. We recognize that extreme clustering of fossil remains precludes adequate testing of spatial hypotheses in some cases. We suspect, however, that there are many cases for which the fossil record permits these kinds of questions to be answered.

We selected *Archaeolemur* (a subfossil indrioid from Madagascar) as an appropriate subject for our study of spatial variation in habitat and morphology. Madagascar has a rich fossil primate fauna from abundant localities, especially in the south, southwest coastal, and central regions (all west of the eastern rainforest). Remains of *Archaeolemur, Pachylemur, Megaladapis* and *Palaeopropithecus* are broadly distributed over the island. (Figure 1 represents only those sites which have yielded remains of the archaeolemurines.) Although all of these forms became extinct recently, apparently as a result of the recent arrival of humans on Madagascar and the subsequent deforestation (Walker 1967a, Mahé 1972), the distribution reflected in the subfossil record most probably preserves the pattern manifested prior to the arrival of people.

Size/shape variation in *Archaeolemur* can be studied in a "single time slice"; samples of *Archaeolemur* are geologically contemporaneous, radiocarbon dates indicate an extremely recent extermination, and no subfossil sites on Madagascar have been found to date back more than several thousand years B.P. (Walker 1967a, Mahé 1972, Battistini and Vérin 1967, Tattersall 1973). Temporal variation, therefore, does not confound the question of spatial variation. Sample sizes are small (though respectable for fossil primates). Reconstructing paleoenvironment is less a problem than might be anticipated. Since the time of extinction there has been little change in climate or geology (Donque 1972, Battistini 1972). Even though substantial degradation of the vegetation has taken place, 80% or more of species are endemic (Paulian 1961, Koechlin 1972, Chauvet 1972). The degradation appears to be due to human slash and burn agriculture followed by rapid soil erosion rather than any *substantial* change in climate; indeed, the climate supports remnant patches of endemic vegetation everywhere. For these reasons, data on "primary phytogeographic zones", "mean annual rainfall", "temperature", (and, of course, "altitude") collected today can be expected to closely reflect conditions which prevailed when *Archaeolemur* lived.

The primary goals of our study of *Archaeolemur* were to isolate scale-related changes in shape and explore external factors which may have influenced the evolution and maintenance of the observed morphoclines. A secondary goal, to describe the nature of the size cline within the context of genetic models which might account for the changes in adult morphology (changes in the genetic basis of morphogenesis or developmental processes), will be treated in a separate paper.

Remains of archaeolemurines (*Archaeolemur* and the closely related but sparsely represented *Hadropithecus*) have been recovered from some 20 sites on Madagascar (Fig. 1, Table 1). For any variable, or any single bone, sample sizes never exceeded 96, with subsamples of 1 to 30 specimens from any single locality. More problematic was the uneven (clustered) distribution of specimens. Only a few sites (Andrahomana from Southeast Madagascar, near Fort Dauphin, Beloha Anavoha from Southwest Madagascar, and Ampasambazimba from Central Madagascar) were well represented. Given data of this sort, how can one convincingly *demonstrate* a cline?

Indeed, variation in *Archaeolemur* has not been previously described as clinal. Instead, two species have been named, each based on materials from two highly

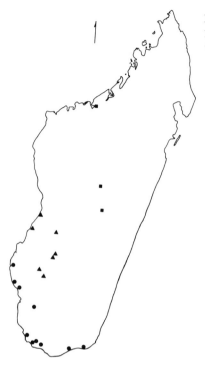

Fig. 1. Map of Madagascar showing locations of sites yielding *Archaeolemur*. *Squares, triangles,* and *circles* depict geographic zones

Table 1. Sites yielding specimens of *Archaeolemur* used in this study

North coastal and central ("Central")	West central coastal and south central ("South Central")	Southwest coastal and southern ("South")
Ampasambazimba	Ambararata Mahabo	Ambolisatra
Ampihingidro	Ampoza	Ambovombe
Antsirabé	Ankazoabo	Anavoha
	Belo	Andrahomana
	Tsiravé	Beloha
	Tsitondroina	Bemafandry
		Betioky Tulear (Taolambiby)
		Bevoha
		Itampolové
		Lamboharana
		Manombo Tulear

fossiliferous regions — the larger *Archaeolemur edwardsi* from several sites in Central Madagascar; the smaller *Archaeolemur majori* from southern and southwest coastal sites. Differences in scale in *Archaeolemur* are well known (Lamberton 1937, Jouffroy 1963, Walker 1967b, Tattersall 1973; Godfrey 1977). Tattersall (1973) attributes the major differences in cranial form of the two species *A. edwardsi* and *A. majori* to differences in size. For example, he described hypertrophication of M. temporalis in

A. edwardsi leading to the presence of a sagittal crest in this species where none exists in *A. majori*. Not too surprisingly, some taxonomic confusion has plagued placement of some specimens from the more *poorly represented* geographically intermediate zone (West Central coastal and South Central Madagascar). For instance, Alan Walker (1976b) working at the British Museum of Natural History, based his description of *Archaeolemur edwardsi* solely upon a fairly complete and robust specimen of *Archaeolemur* from Ampoza in South Central Madagascar. Unfortunately, although the skeleton was remarkably undamaged and well-preserved (many epiphyses, some hand bones and vertebrae preserved), there was no associated *skull* in the collections of the British Museum. It would be surprising indeed if there was no associated skull at the time of discovery. Indeed, there may well have been one, and it may have been sent to the American Museum of Natural History by its collectors. The collecting expedition was the "White-Delacour" mission, a joint enterprise of bone hunters from New York, Paris, and London, which is known to have divided associated skeletal elements among participants. In any case, there is a skull in the New York Museum which was collected by the same expedition from the same geographic area of Madagascar, which is of a similar developmental age and in a similar state of excellent preservation. This skull was included by Ian Tattersall (1973) in his group of typical *A. majori* [3]. If this skull of *Archaeolemur majori* and postcranial skeleton of *Archaeolemur edwardsi* are from the same individual, we have an obvious taxonomic "problem". Of course, two species of *Archaeolemur could* have inhabited Ampoza, and if we were to believe the various early descriptions of specimens from this general region, we might come to that conclusion. To state, however, that the differences in size are *clinal* implies that, whatever the level of genetic differentiation, the differences are less abrupt than would be expected otherwise, falling in a pattern of continuous or stepped change along a geographic transect or transects. Given a size cline, one would be tempted to examine the pattern of variation in a number of ecological variables which might control the cline. Thus we have a serious research problem. The clinal interpretation should be subjected to rigorous testing.

Testing the Clinal Hypothesis

A variety of mathematical tools can be used to test the clinal hypothesis: a list of the techniques we used, the *expected* results for clinal size distribution, and the *observed* results for *Archaeolemur,* are presented in Table 2. We are preparing a detailed analysis of the expectations of five distinct hypotheses (Petto and Godfrey, in prep.). Only the clinal hypothesis will be considered here.

---

3 Actually, Tattersall (1973) contradicts himself in listing *A. edwardsi* only from Ampoza on p. 13, and *A. majori* only from Ampoza in his map and discussion of the distribution and taxonomy of *Archaeolemur* on pp. 24–25. The latter was apparently based on his assessment of the cranial materials from the region; the former, apparently, on Walker (1976b). The skull specimen in question (American Museum of Natural History #30007) is catalogued as *A. majori*. The postcranial materials in question are M. 13924–13929 (British Museum of Natural History). Tattersall acknowledges taxonomic uncertainty only at Amparihingidro (near Majunga) and Ambolisatra (near Manombo and Tulear)

The first requisite in demonstrating a morphocline is that significant geographic differences in form exist (Table 2, Col. 1). Measurements recorded for 73 crania and upper dentitions, 96 mandibles, 51 humeri, 34 tibiae, and 18 radii of *Archaeolemur* were subjected to one-way analysis of variance. (No other bones were present in sufficient quantities to warrant size distribution analysis.) Since specimens were often damaged or incomplete, univariate analysis must be used if the largest available sample sizes are to be obtained. Furthermore, the bones measured were rarely derived from associated skeletons. For each analysis, then, the actual "sample" of individuals drawn varied. The consistent pattern which emerged could not be an artifact of sampling a small number of individuals.

More than 200 craniodental and postcranial variables were tested for significant between-site (and between "grouped-sites") variation, using one-way analysis of variance. The null hypothesis (no difference between sites or grouped sites) was easily rejected for the great majority of variables. This establishes geographic differences but does not address the question of *pattern*. To do so, we grouped the 20 sites bearing remains of *Archaeolemur* into three contiguous geographic "zones". Once defined, zone can be rank-ordered and treated as an ordinal variable. It goes without saying that the zones must be definable without reference to the specimens contained within them to avoid a circular demonstration of a size cline. The criteria we used were geographic and ecological; the sites were grouped into zones reflecting a spatial and ecological gradient. These zones are represented in Fig. 1, a map depicting fossil sites, as solid squares, triangles and circles:

1. The squares represent "north coastal and central" sites. The northwest coastal region is characterized by mangrove and tropophile forest; the central sites (including the famous site Ampasambazimba) occupy a plateau area of (formerly) relatively moist deciduous forest.

2. Triangles represent "west-central coastal and south central" sites. This is a much drier region characterized by more open deciduous forest.

3. Circles depict "southwest coastal and southern" sites. West of Fort Dauphin at the southern end of the east coast, a sharp transition occurs to a xerophilous vegetation consisting largely of scrub. This dry, relatively open region extends westward and northward up the southwest coast.

These zones are not sharply demarcated but reflect ecoclines in altitude, rainfall, and vegetation. The subfossil sites, all of which dot the area west of the tropical rainforests along the eastern coast of Madagascar, are differentially affected by the rain-bearing trade winds moving in from the east. The eastern rainforests give way to deciduous forests reflecting progressively drier climate from east to west, and from north to south. The vegetational ecocline is today somewhat marred by human habitat destruction and erosion, but scattered remnants of native forest in devastated regions record clinal differences in vegetation which existed at the time *Archaeolemur* and other extinct lemurs flourished. In sum, sites bearing the remains of *Archaeolemur* can be grouped along a geographical ecological cline controlled by climatic and topographic variables.

The "test of linearity" (Nie et al. 1975, p. 257; Table 2, Col. 2) makes explicit use of zone as an ordinal variable. This test breaks "between-group" variance into two portions – a sum of squares associated with a linear regression calculated for "zone"

Table 2. Testing the clinal hypothesis for changes in scale

	Anova (1)	Test of linearity (2)	Test of rank order association (3)	Bivariate regression analysis (4)	R-mode factor analysis (5)	Contour mapping (6)
Expected of clinal variation	Significant between-zone differences	No significant deviations from linearity	Gamma tests of rank order association between geographic "zone" and zone means for other parameters demonstrate strong association	One "way of varying"	"Size" factor axes isolate variance along a geographic cline. (Can apply gamma test of rank order association)	Contour intervals for mean site scores on "size" factor axes confirm clinal variation
Observed check indicates expectations confirmed for *Archaeolemur*	✓	✓	✓	✓	✓	✓

Table 3a. Cranial and dental data, N = 77. 75 variables showing significant[a] between- vs within-zone variance

Zone	Smallest	Intermediate	Largest
South	67	8	0
South central	8	61	6
Central	0	6	69

a Anova F test of significance: P < 0.01

Table 3b. Humeri, N = 51. Ten variables showing significant[a] between- vs within-site variance

Zone	Smallest	Intermediate	Largest
South	9	1	0
South central	1	9	0
Central	0	0	10

a Anova F test of significance: P < 0.05

vs "variable X", and a sum of squares associated with *deviations* from the linear regression. It then compares the deviations sum of squares to the pattern of variation observed *within* geographic zones (within-group sum of squares). In the case of *Archaeolemur,* for variable after variable, most of the between-group variance did indeed fit a linear regression model; deviations from that model were trivial.

A second test of the clinal hypothesis might be devised by examining the rank-order of variable means (Table 2, Col. 3). Table 3 shows several examples of the pattern which emerges when the rank orders of variable means for zones are tabulated. We used only those variables previously shown to exhibit significant geographic variation (we used the 0.01 level of confidence for craniodental data, and because smaller sample sizes mandated looser standards, the 0.05 level of confidence for postcranial data). For each variable, the zones with the lowest, intermediate, and highest means were marked. Table 3a shows the results for maxillary and skull data (75 significant variables); Table 3b shows the results for humeri (10 significant variables). An obvious size cline, running geographically from the southern and southwestern portion of Madagascar to the north-central zone, is indicated by this nonrandom pattern. (If necessary, the strength of the rank order association can be tested using the gamma statistic. Gamma varies from + 1 to − 1 and measures the amount of concordance and discordance in rank order pairs. Values close to + 1 or − 1 indicate high association; the sign describes the ordering. Here it is obvious that the association is very strong.)

Postcranial and cranial variables reveal the same geographically controlled size cline. However, because postcranial sample sizes tended to be small, many variables lacked F's significant to the $P < 0.05$ level. It is noteworthy that the same rank order of means emerged for many variables whose between-group sums of squares were not significantly greater than their within-group sums of squares. Given even the poor samples of a fossil population, a highly nonrandom pattern emerges from these data.

More powerful tests of the clinal hypothesis can be devised using techniques which do not presume a priori groupings. Artificially constructed groups may mask natural patterns of change (gradual, abrupt). Furthermore, it is not at all clear whether our assigned zones best reflect change in vegetation, rainfall, altitude, morphology, etc., or just how these variables interrelate. Therefore we must consider methods of analysis which do not *begin* with groups defined by the investigator, and which can elucidate interrelationships among variables with known spatial coordinates.

Bivariate regression analysis (Table 2, Col. 4) does not begin with groups defined by the investigator. Scattergrams depict "ways of varying"; more than one "way of varying" suggests an abrupt shift in adaptive strategy. For clinal change, our expectations are that there will be only one relationship reflected in the data, and that the fit of cases from all geographic zones to the least squares solution will be good. Morphological variation in *Archaeolemur* was analyzed in this manner. We found no example of abrupt change in relationship. Our further expectation − that individuals from the intermediate geographic zone should consistently fall in the middle range of the scattergram − was also corroborated by the data (cf. Fig. 2, for an example).

Of course, bivariate regression analysis can be exceedingly redundant and tedious when applied to a mass of metric data. Ways of varying are better studied using multivariate data reduction techniques, such as R-mode factor analysis (Hursh 1976).

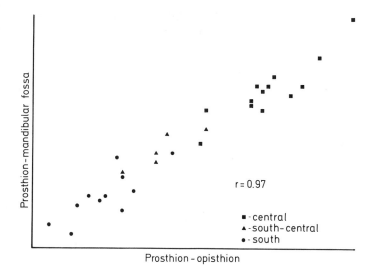

Fig. 2. Sample bivariate regression analysis

The factor axes isolate identifiable patterns of variation, including scale-related changes in form. Factor analysis minimizes redundancy, and treats patterns of change in morphocomplexes as single variables. Variable *loadings* elucidate the biological meaning of a particular factor axis and tell the investigator how much of the total variance is "explained". The *geography* of change in scale can be studied by examining case scores (i.e., factor scores for individual fossils from different sites)[4]. By examining *both* factor loadings and factor scores the investigator can explore the *nature* of the change isolated by the factor axis, and the geographic *pattern* of that change.

Still we need a way of describing spatial variation as a continuous phenomenon on a multidimensional "surface". Trend surface analysis fulfills our major objectives: it begins without a priori groupings and it treats spatial variation as a multidimensional problem, generating an equation for a surface which best fits the observed spatial variation, and which can be solved for any X and Y coordinate on that surface. However, it makes some bold assumptions and therefore has some limitations; it will be considered after a discussion of simpler mapping techniques, which can also be used to test a clinal hypothesis.

We began with SYMAP contour mapping (Dougenik and Sheehan 1975). It divides the total range of measurements into a specified number of class intervals of equal size, and then plots contours encompassing all sites falling within these specified intervals; it therefore eliminates observer bias in constructing "zones" based on any single criterion. Contour mapping provides a clear graphical representation of the variation (clinal or otherwise) in any parameter. It provides a visual test of the clinal hypothesis (Fig. 3), since it is clear that the width and distribution of the contour bands describe the pattern of variation.

---

4 We used SPSS principal factoring without iteration (PA1), and obtained factor scores using the Varimax and Oblique rotations

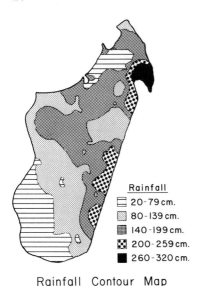

Fig. 3. Facsimile of Madagascar generated using SYMAP mapping procedure. Rainfall contours based on five intervals

Rainfall Contour Map

The clinal nature of the "size" factor axes (generally the first factor axis of each run) was ascertained by calculating the *mean factor scores* for individuals from the different sites and mapping them. (SYMAP cannot handle more than one value *for any particular site.*) If, for example, a particular factor analysis was based on a sample of 71 individuals from 12 sites, we calculated 12 site means from the factor scores for the 71 individuals. We felt justified in calculating factor scores for sites, because analysis of variance had already indicated significant between-site and between-zone differences, especially for variables which best reflect scale.

Given the fragmentary nature of fossil remains, it was impossible to run a comprehensive factor analysis on the complete skeleton or even major portions of the skeleton. Instead, we ran separate analyses on functionally integrated *portions* of bones (e.g., the cheek teeth, the face, the neurocranium, the body of the jaw, etc.), thereby maximizing sample sizes while treating structural-functional complexes as completely as possible. Even so, the largest sample available was indeed 71 mandibular cheek tooth rows from 12 sites. Runs based on samples of 25, 45, or 60 specimens were more common. The number of sites represented in any single run was usually half of the 20 total sites bearing remains of *Archaeolemur.* (Six of these were common to most of the analyses.) In the end, we obtained a dozen factor analyses which provided us with raw material for contour mapping.

Some contour maps generated from these analyses are shown in Fig. 4a,b,c. Each map describes the variation represented by a single factor axis. The dark zones include sites with high mean factor scores; the lightest zones depict low scoring sites. Each map is a "multivariate" picture of clinal size variation in *Archaeolemur* on Madagascar. The consistent nature of the clinal variation depicted by these contour maps is striking. Note that they were produced by different factor analytic runs, based on different variables.

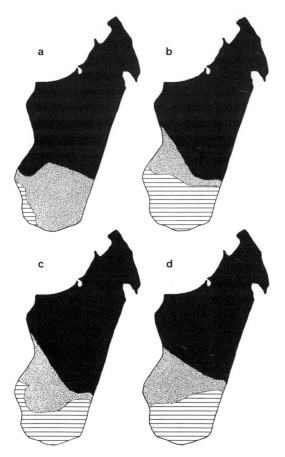

Fig. 4 a–d. Size axes generated from factor analyses of craniofacial data: a Craniofacial N = 22, 10 sites. b Maxillary cheek teeth N = 45, 12 sites. c Mandibular cheek teeth N = 50, 11 sites. d Summary size

We also generated contour maps for climatic, topographic, and ecological data taken from published sources (Battistini 1864, Chauvet 1972, Donque 1972, Koechlin 1972, Paulian 1961; Tattersall and Sussman 1975; and the Times Atlas). Our "mean annual rainfall" map (Fig. 3), for example, was based on values for 98 sites; our maps of "vegetation zones"[5] and "altitude" were based on 174 sites.

We could now demonstrate *not only* that we had a series of size-based morphoclines, but that these morphoclines are associated with ecoclines in such variables as mean annual rainfall and vegetation. We used rank order data provided by the SYMAP contour mapping program (computer-generated rainfall zones, vegetation zones, etc.) and compared site values using the gamma test of rank order association. Several examples are given in Tables 4 and 5. Here we compare the rank order for sites in our original assigned "geographic zones" with their rank order in rainfall and vegetation zones generated by the computer. The values for gamma are very high; a strong association is indicated.

---

5   For concise definitions of the zones we employed, see Tattersall and Sussman (1975)

Table 4. Site value (contour zone) for vegetation

Zone	1	2	3	4	5
Central	0	1	2	0	0
South central	0	8	0	0	0
South	9	1	0	0	0

$\gamma = -1.0$

Table 5. Site value (contour zone) for rainfall (in cm)

Zone	20–79	80–139	140–199	200–259	260–320
Central	0	0	3	0	0
South central	3	4	0	0	0
South	9	1	0	0	0

$\gamma = -0.9333$

We could, furthermore, statistically demonstrate that our original geographic zones largely matched SYMAP-generated contour zones based on factor axes describing variation in "size". We ran gamma tests of rank order association between our *assigned* site zone values and the contour values constructed by SYMAP for our set of *first* factor axes (isolating variance in size and allometrically related changes in shape), *second* factor axes, and *third* factor axes (Table 6). The value of gamma for the set of first factor axes was very high, less so for the set of second factor axes, and low for the set of third factor axes. It is clear that most of the variation in form which fits the geographic cline we originally defined is related to change in scale.

SYMAP trend surface maps can be generated from the same data used to construct contour maps. Trend surface analysis produces an equation which describes the cline as a spatial phenomenon. It imposes a consistent direction on the data (the surface, a third order polynomial expression, flows from highest to lowest values). Therefore it can only be justified when a cline (with extremes) has already been demonstrated. The surface is not fitted to the exact values of the data points provided by the investigator; it is, rather, a least squares solution. The main advantage of trend surface analysis is that it allows one to extract values for any site on the map. We might caution, however, that the equation produced represents an extrapolation from actual data which can err markedly in regions which have not been properly sampled. SYMAP also provides a map of residuals which allows the investigator to judge the fit of the equation to observed values. We felt justified in using trend surface analysis (but then only for the region of Madagascar dotted by fossil sites) because:

1. we had already demonstrated clinal variation to our satisfaction; and
2. the residuals in fossil-bearing regions for all maps were very low.

Figure 5 shows the rainfall trend surface map and a map of rainfall residuals.

Now it was possible to examine associations between any isolated morphocline and any isolated ecocline using the familiar parametric techniques of correlation and

Table 6a. Site value (contour zone) for first factor axes

Zone	Low	Intermediate	High
Central	0	1	15
South central	8	12	2
South	22	4	0

$\gamma = -0.93645$

Table 6b. Site value (contour zone) for second factor axes

Zone	Lowest	Low	Intermediate	High	Highest
Central	1	0	1	1	5
South central	2	2	5	0	2
South	11	2	3	0	0

$\gamma = -0.7875$

Table 6c. Site value (contour zone) for third factor axes

Zone	Low	Intermediate	High
Central	0	3	2
South central	0	4	2
South	3	2	3

$\gamma = -0.4286$

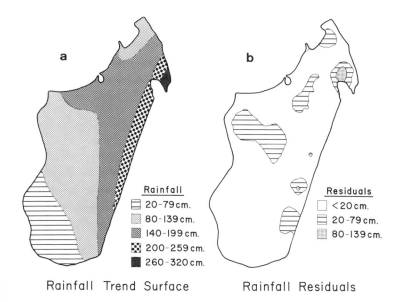

Rainfall
☐ 20-79 cm.
▨ 80-139 cm.
▨ 140-199 cm.
▨ 200-259 cm.
■ 260-320 cm.

Residuals
☐ <20 cm.
☐ 20-79 cm.
▨ 80-139 cm.

Rainfall Trend Surface   Rainfall Residuals

Fig. 5a,b. SYMAP trend surface and residual maps for rainfall

and regression. One example will suffice to demonstrate the power of this technique. We constructed a "Summary Size" factor axis by taking size factors from different runs as *variables*, the six common sites as *cases*, their mean scores on reach variable as *values* for the variables. These data were treated by factor analysis to isolate *common* variance in size. A new set of site values (factor scores), derived from the first axis in the oblique solution[6], was used to construct a Summary Size contour map (Fig. 4d). The same six values were used to construct a Summary Size trend surface equation and a map of residuals. Our six values fit the trend surface equation perfectly (Table 7).

Table 7. Residuals for Summary Size. (Predicted and observed values of factor scores for six subfossil localities)

Site	Observed mean factor score (Summary Size) (Z-value)	Predicted score	Residual
Bemafandry	− 0.907	− 0.907	0.000
Beloha Anavoha	− 0.861	− 0.861	− 0.000
Andrahomana	− 0.572	− 0.572	0.000
Ampoza	0.222	0.222	0.000
Tsirave	0.420	0.420	0.000
Ampasambazimba	1.698	1.698	0.000

Our use of 6 to 12 values for the construction of any contour or trend surface map describing morphological variance in *Archaeolemur* was dictated by the fragmentary nature of fossil remains. While, due to these extremely small samples, we must interpret our results with caution, we feel that our observed relationships are "real". They will be treated in more detail in a future publication; here we will briefly outline some of the results. (Of course the *technique* is a valid and useful one; although its power in elucidating relationships depends upon its sample base.)

We selected 74 sites from portions of Madagascar which have yielded subfossil remains (south, southwest, south central, central, and north central regions), and compiled trend surface scores for Summary Size, rainfall, altitude, and vegetation. (Regions of Madagascar eliminated from consideration are shown in Fig. 6.) Now we were ready to examine the ecological context of change in size in some detail. (Note that we used, not *observed*, but *predicted* values for each of the variables considered in this study.)

Table 8 shows the Pearson coefficient of correlation for all variable pairs based on our sample of 74 sites. Because we eliminated the eastern rainforest from consideration, these results are not confounded by the very different relationships between variables (such as rainfall and altitude) which exist east and west of the central mountain range. However, the positive relationship between rainfall and altitude *was* partly

6 When case vectors are tightly clustered, the first oblique axis generally explains more variance than does the first varimax axis; it approximates the first principal component which is the best description of size. Before using factor scores generated by this oblique solution, we checked the variable loadings to be sure this was the case

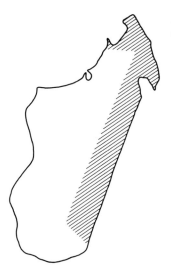

Fig. 6. Regions of Madagascar eliminated from correlation and regression analyses

Table 8. Pearson's r for trend surface scores (N = 74)

Variable pair	r	Significance
Rainfall and vegetation	0.82	P ≤ 0.0001
Vegetation and altitude	0.80	P ≤ 0.0001
Rainfall and altitude	0.51	P ≤ 0.0001
Rainfall and summary size	0.42	P ≤ 0.001
Vegetation and summary size	0.08	NS
Altitude and summary size	− 0.10	NS

confounded by other geographic trends in temperature and moisture (north-south, east-west). The northern coast of Madgascar is moist; the southern coast is very arid, and the area near Ft. Dauphin marks a sharp transition from very arid to very wet.

Of the variables we examined, rainfall was the best predictor of change in scale in *Archaeolemur*. Size increases with *increasing* rainfall. The correlation between "mean annual rainfall" and "summary size" was highly significant (Table 8); rainfall was also positively correlated with "vegetation" (increasingly lush vegetation with increasing rainfall) and "altitude" (more rainfall at higher altitudes).

Vegetation was a weak measure of changes in flora. The trend surface for this variable was based on an original ranking of primary phytogeographical domains (cf. Humbert 1955, Tattersall and Sussman 1975) from highly xerophilic (zone 1) to the high mountain rainforest domain (zone 5). Of the original five vegetation zones defined, our sample of 74 sites was drawn only from the first three. Therefore vegetation (as used here) is a less powerful index of ecological variation than rainfall. An examination of partial correlation coefficients reveals that, *within constant vegetation zones,* size increases with increasing rainfall (the partial correlation of rainfall with summary size, holding vegetation constant, was 0.61, P ≤ 0.001). In zones of *constant elevation,* size increases with increasing rainfall (P ≤ 0.001) and with increased lushness of vegetation (P ≤ 0.02).

The variance in size which is positively associated with change in vegetation was better explained by increased rainfall. However, there does seem to be a coastal equability factor which is not reflected in either the rainfall or vegetation parameters employed here. With rainfall *held constant,* size increases toward the coast (size is inversely correlated with altitude) ($P \leqslant 0.001$). We caution that very little of the variance in size is being explained by this apparent coastal equability factor; its real importance (or even, validity) requires investigation of ecological variables not considered here – measures of temporal stability and seasonal variance. Given the small amount of variance in size explained by this partial correlation, we cannot be sure that this "effect" is not an artifact of extrapolating a "trend" based on six sites to coastal regions for which the trend is not entirely valid.

Multiple regression analysis (taking summary size as dependent variable, and rainfall, vegetation, and altitude as independent variables) reaffirmed the importance of rainfall as a predictor of size. The multiple R (0.611) was highly significant ($P \leqslant 0.001$; df = 70). The regression also had a highly significant F ($P \leqslant 0.001$). Both vegetation and rainfall added significantly to the regression. The inclusion of altitude had a negligible effect upon variance in size explained by the regression.

Summary and Conclusions

The ecological context of size variation in congeneric species has been a subject of great interest in the past decade (cf., for example, Geist 1971, on mountain sheep; Grubb 1972, on water buffalo; Shvarts 1975, on voles; and Heaney 1978, on tricolored squirrels). We have examined the ecological context of size variation in the extinct indrioid *Archaeolemur.* In this case we see progessively larger variants of *Archaeolemur* occupying wetter, lusher habitats. Such associations have been noted by previous researchers. Gould (1977) noted a tendency in "k-selective regimes" for selection to favor hypermorphosis, size increase, and delayed sexual maturation (p. 341). Such selection will depend upon the abundance of potentially limiting resources. In resource-limited environments, *small* size will be favored (Sondaar 1977, Wassersug et al. 1979). Wassersug et al. (1979) have argued that stochastic processes alone will favor decrease in size in predator free but biomass-limited, resource-limited, environments (conditions common in insular situations): "given that there are temporal fluctuations in the available resources, dwarfs should be able to utilize the available resources more completely than giants and, at the same time, are less likely to exceed the environmental carrying capacity and crash" (p. 292). Such changes in size should permit the "closest tracking of carrying capacity when immigration and emigration are not possible" (p. 292). That such factors *could* have controlled the size/shape gradient in subfossil populations of *Archaeolemur* seems evident. However, detailed analysis must await treatment of the nature of the observed size clines and consideration of genetic models which might account for the spatial changes in adult morphology. Here we have concentrated our efforts on showing that it is possible to test hypotheses of spatial orientation in fossil forms and establish an ecological context to change in form (even when sampling problems exist).

Despite the frequent application of mapping techniques, such as trend surface analysis, to problems of spatial orientation in other fields (geography, geology; cf., for example, Lee 1969), trend surface analysis has rarely been employed in testing such hypotheses in primatology or biological anthropology. Its promise as a research tool for the study of variation in extinct forms is only *partly* hampered by frequent uneven clustering of fossil sites and small sample sizes. It is true that a map extrapolated from observations on 6 sites cannot be treated with the respect which a map extrapolated from 600 deserves. Furthermore, trend surface analysis should not be used as a preliminary test of clinal variation since the method will produce a smoothly flowing surface *whatever* the actual data look like. It is circular to use it to *demonstrate* a cline. (More appropriate to demonstrate clinal variation is contour mapping, used in conjunction with various means of testing for rank order association.) But, if employed with these considerations in mind, trend surface analysis can be a valuable tool for elucidating spatial relationships, especially when combined with other means of isolating patterns of variation in whole morphocomplexes.

It has been long recognized that factor analysis is a highly useful and appropriate means to isolate scale-related change in shape. "Size" axes can be used to elucidate patterns of scale-related change through an examination of variable loadings. They can also provide new material for analysis of spatial variation (in factor scores of individuals, or, for purposes of mapping, mean scores of individuals from fossil localities).

Acknowledgments. We owe our sincerest thanks to the many people who helped in various stages of this research, either with technical assistance, making specimens available for study, or hours of fruitful discussion. Specimens were made available by: Miss Theya Molleson, British Musem of Natural History; Dr. Roger Saban, Dr. J. Lessertisseur, and Mme. F.K. Jouffroy of the Laboratoire d'Anatomie Comparée, Paris; Dr.J.P. Lehman, Dr. J. Brunet, and Dr. J. Mahé, Institut de Paléontologie, Muséum National d'Histoire Naturelle, Paris; Dr. H. Kollmann, Naturhistorisches Museum, Geologisch-Palaeontologische Abteilung, Vienna; the late Dr. P. Radaody-Ralarosy, Académie Malgache, Tananarive; Dr. P. Roederer, Centre O.R.S.T.O.M. de Tananarive; Dr. P. Vérin, Laboratoire de Géographie et Centre Archéologie, Université de Tananarive; Dr. Sydney Anderson and Dr. Fred Szalay, American Museum of Natural History, New York.

Artwork for this manuscript was done by Larry Gallant and Paul J. Godfrey. We would also like to thank Susan Hancock who aided with the computer analyses of craniofacial data, H.M. Wobst and J.A. Hafner for advice on SYMAP analysis, and D. Weber-Burdin who was instrumental in bringing each data run to fruition.

This research was supported in part by National Science Foundation Grant #GS-3226 and a grant from the University Computing Center at the University of Massachusetts, Amherst.

References

Alexander R McN, Goldspink G (1977) Mechanics and energetics of animal locomotion. Chapman & Hall, London New York

Battistini R (1964) Étude géomorphologique de l'extrême sud de Madagascar. Editions Cujas, Paris

Battistini R (1972) Madagascar relief and main types of landscape. In: Battistini R, Richard-Vindard G (eds) Biogeography and ecology in Madagascar. Junk, The Hague, pp 1–25

Battistini R, Vérin P (1967) Ecologic changes in protohistoric Madagascar. In: Martin PS, Wright HE Jr (eds) Pleistocene extinctions: The search for a cause. Yale University Press, New Haven London, pp 407–424

Bock WJ, von Wahlert G (1965) Adaptation and the form-function complex. Evolution 19:269–299

Bonner JT (1965) Size and cycle. Princeton Univ Press, Princeton NJ

Bonner JT (1974) On development. Harvard Univ Press, Cambridge Ma

Bourlière F (1975) Mammals, small and large: The ecological implications of size. In: Golley FB, Petrusewicz K, Ryszkowski L (eds) Small mammals: Their productivity and population dynamics. Cambridge Univ Press, Cambridge New York

Brody S (1945) Bioenergetics and growth; with special reference to the efficiency complex in domestic animals Reinhold, New York

Carson HL(1973) Reorganization of the gene pool during speciation. In: Morton NE (ed) Genetic structure of populations. Popul Genet Monogr 3:274–280

Cartmill M (1974) Pads and claws in arboreal locomotion. In: Jenkins FA Jr (ed) Primate locomotion. Academic Press London New York, pp 45–83

Charles-Dominique P (1975) Nocturnality and diurnality; an ecological interpretation of these two modes of life by an analysis of the higher vertebrate fauna in tropical forest ecosystems. In: Luckett WP, Szalay FS (eds) Phylogeny of the primates: A multidisciplinary approach. Plenum Press, New York London, pp 69–88

Chauvet B (1972) The forest of Madagascar. In: Battistini R, Richard-Vindard G (eds) Biogeography and ecology in Madagascar. Junk, The Hague, pp 191–199

Clutton-Brock TH, Harvey PH (1977) Species differences in feeding and ranging behaviour in primates. In: Clutton-Brock TH (ed) Primate ecology: Studies of feeding and ranging behaviour in lemurs, monkeys and apes. Academic Press, London New York San Francisco, pp 557–584

Donque G (1972) The climatology of Madagascar. In: Battistini R, Richard-Vindard G (eds) Biogeography and ecology in Madagascar. Junk, The Hague, pp 87–144

Dougenik JA, Sheehan DE (1975) SYMAP user's reference manual: Laboratory for computer graphics and spatial analysis. Graduate School of Design, Harvard Univ, Cambridge Ma

Eldredge N, Gould SJ (1972) Punctuated equilibria: An alternative to phyletic gradualism. In: Schopf TJM (ed) Models in paleobiology. Freeman, Cooper and Co, San Francisco

Endler JA (1977) Geographic variation, speciation and clines. Princeton Univ Press, Princeton NJ

Fleagle JG (1979) Primate positional behavior and anatomy: naturalistic and experimental approaches. In: Morbeck ME, Preuschoft H, Gomberg N (eds) Environment, behavior, and morphology: Dynamic interactions in primates. G Fischer, New York Stuttgart, pp 313–325

Frazzetta TH (1975) Complex adaptations in evolving populations. Sinauer Associates, Sunderland Ma

Geist V (1971) Mountain sheep: A study in behavior and evolution. Univ of Chicago Press, Chicago London

Godfrey LR (1977) Structure and function in *Archaeolemur* and *Hadropithecus*. Ph D Diss, Harvard Univ

Gottlieb LD (1975) Biochemical consequences of speciation in plants. In: Ayala FJ (ed) Molecular evolution. Sinauer Associates, Sunderland Ma, pp 123–140

Gould SJ (1966) Allometry and size in ontogeny and phylogeny. Biol Rev 41:587–640

Gould SJ (1970) Evolutionary paleontology and the science of form. Earth Sci Rev 6:77–119

Gould SJ (1971) Geometric scaling in allometric growth: A contribution to the problem of scaling in the evolution of size. Am Nat 105:113–136

Gould SJ (1975) Allometry in primates, with emphasis on scaling and the evolution of the brain. Contr Primatol 5:244–292

Gould SJ (1977) Ontogeny and phylogeny. Belknap, Cambridge Ma London

Gould SJ (1980) Is a new and general theory of evolution emerging? Paleobiology 6:119–130

Gould SJ, Eldredge N (1977) Punctuated equilibria: The tempo and mode of evolution reconsidered. Paleobiology 3:115–151

Grubb P (1972) Variation and incipient speciation in the African buffalo. Z Säugetierkd 37: 121–144

Hairston HG, Tinkle DW, Wilbur NM (1970) Natural selection and the parameters of population growth. J. Wildl Manage 34:681–690

Heaney LR (1978) Island area and body size of insular mammals; evidence from the tri-colored squirrel *(Callosciurus pevosti)* of Southwest Asia. Evolution 32:29–44

Humbert H (1955) Les Territoires phytogéographiques de Madagascar, leur cartographie. Colloq Int Centre Nat Rech Sci LIX:195–204

Hursh TM (1976) The study of cranial form: Measurement techniques and analytical methods. In: Giles E, Friedlaender JS (eds) The measures of man: Methodologies in biological anthropology. Peabody Museum Press, Cambridge Ma

Jerison HJ (1973) Evolution of the brain and intelligence. Academic Press, London New York

Jouffroy FK (1963) Contribution à la connaissance du genre *Archaeolemur,* Filhol 1895. Ann Paleontol 49:129–155

Kay RF (1975) The functional adaptations of primate molar teeth. Am J Phys Anthropol 43: 195–216

Koechlin J (1972) Flora and vegetation of Madagascar. In: Battistini R, Richard-Vindard G (eds) Biogeography and ecology in Madagascar. Junk, The Hague, pp 145–190

Kurtén B (1953) On the variation and population dynamics of fossil and recent mammalian populations. Acta Zool Fenn 76:1–122

Lamberton C (1937/1938) Contribution à la connaissance de la faune subfossile de Madagascar. Note III. Les Hadropithèques. Bull Acad Malgache Nouv Ser 20:127–170

Lee PJ (1969) The theory and application of canonical trend surfaces. J Geol 77:303–318

Leutenegger W (1973) Maternal-fetal weight relationships in primates. Folia Primatol 20:280–293

Løvtrup S (1974) Epigenetics: A treatise on theoretical biology. Wiley, London New York

Mahé J (1972) The Malagasy subfossils. In: Battistini R, Richard-Vindard G (eds) Biogeography and ecology in Madagascar. Junk, The Hague, pp 339–365

Martin RD (1975) The bearing of reproductive behavior and ontogeny on strepsirhine phylogeny. In: Luckett WP, Szalay FS (eds) Phylogeny of the primates: A multidisciplinary approach. Plenum Press, New York London, pp 265–297

Martin RD (1979) Phylogenetic aspects of prosimian behavior. In: Doyle GA, Martin RD (eds) The study of prosimian behavior. Academic Press, London New York San Francisco

McMahon TA (1975) Using body size to understand the structural design of animals: Quadrupedal locomotion. J Appl Physiol 39:619–627

Milton K, May ML (1976) Body weight, diet and home range area in primates. Nature (London) 259:459–462

Morbeck ME, Preuschoft H, Gomberg N (1979) Environment, behavior, and morphology: Dynamic interactions in primates. G Fischer, New York Stuttgart

Nie NH, Hull CH, Jenkins JG, Steinbrenner K, Bent DH (1975) Statistical package for the social sciences, 2nd edn. McGraw-Hill, New York St. Louis

Passingham RE (1975) Changes in the size and organization of the brain in man and his ancestors. Brain Behav Evol 11:73–90

Paulian R (1961) La zoogéographie de Madagascar et des îles voisines. Faune de Madagascar XIII. Inst Rech Sci, Tananarive-Tsimbazaza

Pedley TJ (1977) Scale effects in animal locomotion. Academic Press, London New York San Francisco

Petto AJ, Godfrey LR (in prep) An analysis of morphological variation in *Archaeolemur* spp. (ms)

Pianka ER (1970) On r and k selection. Am Nat 104:592–597

Pianka ER (1972) R and k selection or b and d selection? Am Nat 106:581–588

Pilbeam D, Gould SJ (1974) Size and scaling in human evolution. Science 186:892–901

Pirie PL (1978) Allometric scaling in the postcanine dentition with reference to primate diets. Primates 19:583–591

Prost JH (1980) Origin of bipedalism. Am J Phys Anthropol 52:175–189

Ripley S (1979) Environmental grain, niche diversification, and positional behavior in Neogene primates: An evolutionary hypothesis. In: Morbeck ME, Preuschoft H, Gomberg N (eds) Environment, behavior, and morphology. Dynamic interactions in primates. G Fischer, New York Stuttgart, pp 37–74

Rose MD (1974) Postural adaptations in New and Old World monkeys. In: Jenkins FA Jr (ed) Primate locomotion. Academic Press, London New York, pp 202–222

Sacher GA, Staffeldt EF (1974) Relation of gestation time to brain weight for placental mammals: Implications for the theory of vertebrate growth. Am Nat 108:593–615

Schmidt-Nielsen K (1970) Animal physiology, 3rd edn. Prentice-Hall, Englewood Cliffs NJ
Schmidt-Nielsen K (1972) How animals work. Cambridge Univ Press, Cambridge New York
Shvarts SS (1975) Morpho-physiological characteristics as indices of population processes. In:
    Golley FB, Petrusewicz K, Ryszkowski L (eds) Small mammals: Their productivity and
    population dynamics. Cambridge Univ Press, Cambridge New York, pp 129–152
Sondaar PY (1977) Insularity and its effect on mammal evolution. In: Hecht MK, Goody PC,
    Hecht BM (eds) Major patterns of vertebrate evolution. Plenum Press, New York London
Stanley SM (1975) A theory of evolution above the species level. Prc Natl Acad Sci 72:646–650
Stanley SM (1979) Macroevolution: Pattern and process. WH Freeman, San Francisco
Stephan H, Bauchot R (1965) Hirn-Körpergewichtsbeziehungen bei den Halbaffen (Prosimii).
    Acta Zool 46:209–231
Tattersall I (1973) Cranial anatomy of the Archaeolemurinae (Lemuroidea, Primates). Anthropol
    Pap Am Mus Nat Hist 52 (1):1–110
Tattersall I, Sussman RW (1975) Notes on topography, climate and vegetation of Madagascar. In:
    Tattersall I, Sussman RW (eds) Lemur biology. Plenum Press, New York, London, pp 13–21
Taylor CR, Schmidt-Nielsen K, Raab JL (1970) Scaling of the energetic cost of running to body
    size in mammals. Am J Physiol 219:1104–1107
Taylor CR, Caldwell SL, Rowntree VJ (1972) Running up and down hills: Some consequences of
    size. Science 178:1096–1097
Waddington CH (1975) The evolution of an evolutionist. Cornell Univ Press, Ithaca NY
Walker AC (1967a) Patterns of extinction amont the subfossil Madagascan lemuroids. In: Martin
    PS, Wright HE Jr (eds) Pleistocene extinctions: The search for a cause. Yale Univ Press, New
    Haven, pp 425–435
Walker AC (1967b) Locomotor adaptation in recent fossil Madagascar lemurs. PH D Diss, Univ
    London
Wassersug RJ, Yang H, Sepkoski JJ Jr, Raup DM (1979) The evolution of body size on islands:
    A computer simulation. Am Nat 114:287–295
White MJD (1978) Modes of speciation. WH Freeman, San Francisco
Wilson DS (1975) The adequacy of body size as a niche difference. Am Nat 109:769–784

The Anatomy of Growth and Its Relation to Locomotor Capacity in Macaca

T.I. Grand[1]

To the problems of growth I bring the attitudes and methods of an anatomist. In the past, I have compared adult forms of many genera, various size classes, and differing locomotor patterns (Grand 1977a, 1978). Here I concentrate on rhesus macaques *(Macaca mulatta)* of several size classes and levels of locomotor skill.

From the data on tissue composition and the distribution of mass to the segments one can reconstruct the shifting center of mass, the differential growth of tissues and regions, and thus, one grades naturally into locomotor capacity and the development of skill. The integrated motor repertoire of play, rest, and free-ranging movements as well as the comparable patterns in man can be examined. Thus, the development of the locomotor apparatus is both the bottom line for detail and the top line for behavioral perspective. Each of the locomotor stages that I propose is related directly to shifts in tissue composition and the segmental distribution of mass.

At the same time, I make no pretense that these data or my interpretations of them can fully satisfy the growth specialist. The usual investigations are longitudinal, i.e., they follow individuals year after year (Tanner 1969, Kowalski and Guire 1974). The growth spurt, for example, is a sharply accelerated change in weight and length in an individual, but when the weights of many individuals of the same age are averaged, individual peaks are smoothed out and the growth spurt blurs. I used cross-sectional sampling because my data were obtained from dead animals. The animals, when arrayed by increasing body weight, represent increasing age. Similar cross-sectional studies have been made on the growth of muscle groups in domestic animals (Berg 1968, Butterfield and Johnson 1968, Widdowson 1968, Fowler 1968) and in wild birds (O'Connor 1975, 1977, Ricklefs 1968, 1973). Here, then, is a common currency for comparative development and for nutrition with the evolutionary adaptation.

Methods

Detailed methods have evolved for the study of adult form and locomotor adaptation. These data on body segmentation and tissue decomposition (or "fractionation") offer a uniform methodological approach to the analysis of body weight. They are also

1 Oregon Regional Primate Research Center, Beaverton, OR 97006, USA

interconvertible with the results of studies on the human body by anatomists, physical educators, and the Ross "school" of kinanthropometrists.

The basic methods of tissue composition and limb segmentation were presented by Grand (1977a,b). Two additions are the analyses of the head and of the truncal segments as well as the greater refinement in rate of growth. The head is taken off in a multistage operation. The head skin is ringed in a line across the throat and around the occipital area. The nuchal muscles, the muscles surrounding trachea and esophagus, and the atlanto-occipital joints are cut. After the head is removed, it is weighed and the tissues separately dissected out and weighed. By dividing the weight of the head by body weight, the percent weight of the head is calculated in % of Total Body Weight (TBW). By dividing the weight of each tissue by the total weight of the head, the percent weight of that tissue is obtained (% of Total Head Weight or THW).

The segmentation and dissection of the trunk are more complicated. After the head is removed the trunk is frozen, and cut into three subdivisions (thorax, lumbar, and pelvic) along two transverse planes (the anticlinal vertebra and the line of the iliac crests). Each segment is weighed, and the tissue are separately dissected (as in the head) after thawing. Each tissue is also weighed. For the thoracic mass, each shoulder is removed to include overlying skin and the propulsive and stabilizing muscles which clothe the scapula. The neck includes some skin, the cervical extensors and flexors, some viscera, and the cervical vertebrae. These tissue groups are dissected away from the thorax and weighed so as to reconstruct the weight of the shoulder masses and the neck, as independently movable segments. The thorax remains with extensors and ribs enclosing the thoracic viscera. For each region the weight of the segment divided by total weight gives the percentage weight of that segment (% TBW). The weight of a tissue divided by the weight of the segment gives the % of that tissue within the segment.

One aspect of growth is increasing weight, and weight is the baseline in this cross-sectional sample. Correlations between weight and age are discussed in a more detailed analysis, but here I simply demonstrate that some tissues and segments grow slower or faster than others with respect to increasing body weight.

The center of gravity is the basis for all biomechanical analyses of motion (Palmer 1944). To locate the center of gravity (Fig. 7), I used the weight or mass of the segments in two individuals. Correction for linear differences between young and adult was made, but not differences in the location of the centers of mass of each segment. Further, the differences in anteroposterior location between one young animal and another mature one represent the postulated shift within the individual from early to late in life. These data in a product of moments calculation (Miller and Nelson 1973) determine the position of the center of gravity.

Results

The transformation of the macaque body from a 250-g fetus to a 6- to 12-kg adult is presented step by step according to anatomical methods.

The tissue compostion of the 450-g neonate (Fig. 1) includes a brain which is 10%, and muscle which is about 24%, of body weight. In the 1500-g juvenile the brain

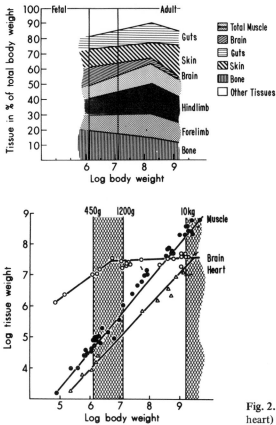

Fig. 1. Changes in tissue proportions from fetus to adult in *Macaca mulatta*

Fig. 2. Growth rates of tissues (brain, muscle, heart) in relation to weight

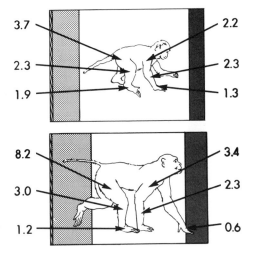

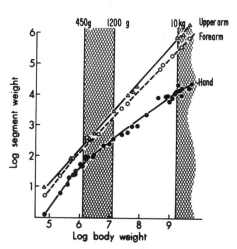

Fig. 3. The relative weight of the limb segments in % Total Body Weight in infant and adult male

Fig. 4. Growth rates of the forelimb segments in relation to weight

drops but muscle increases. In the adult, the brain, overgrown by all other tissues, has dropped to 1% of weight, while muscle represents about 40% of weight. Adipose tissue increases during adulthood.

The brain and eyes (Fig. 2) achieve mature proportions early in life; muscle, heart, and bone grow with locomotor demand and thus account most directly for later body weight increases. Regional increases in the nuchal muscles and the forearm and calf muscles are relatively large early in life; other groups, e.g., shoulder, hip, the masticatory muscles, develop later.

Comparison of the individuals in Fig. 3 demonstrates another feature of the redistribution of weight with increasing size and age. The infant's hand represents 1.5% of weight, the adult's hand about 0.5%. By contrast, at birth the forearm and upper arm are equal, but with age the upper arm becomes heavier and more muscular. The foot drops with age from 2.0% to 1.5%, the thigh doubles to 8%. The forelimb mass rises slightly, but the hindlimbs rise from 16% to 25% of TBW.

The hand, foot, tail, and head are relatively heavy at birth; the thigh and upper arm grow heavier with age (Grand 1977b). Infants have forearm muscles (Fig. 4) that are disproportionately heavier than muscle groups of the upper arm. After 800 to 1,000 g, however, the upper arm muscles grow more.

Truncal segmentation accounts for the remaining 60% to 70% of the body (Fig. 5). In the infant the head represents over 25% of weight, but this declines to 6% to 8% in maturity. The shoulders rise from 3+% of weight to 5+%. The lumbar region rises from 16% or 17% to 20% or 22% of weight. The pelvis rises from 7% or 8% to 11+% of weight. One consistent feature of all these changes is the overall increase in the percentage of musculature.

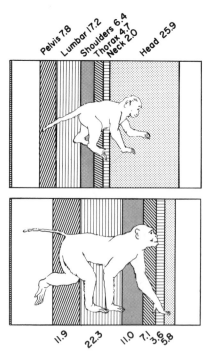

Fig. 5. Segmental distribution of mass within the trunk in infant and adult

The growth of the head shows that despite the systematic drop in head mass, a process of biomechanical importance, the component tissues are changing in still other functional ways. Three changes in composition are significant (Fig. 6). First, the brain achieves full growth by 1000 g. Second, the eyes reach mature size by about 3 kg. Third, at birth, the mandible and masticatory muscles are absolutely smaller than the other tissues. However, as the animal becomes nutritionally independent, as the teeth grow, and as the mandible is subjected to the compressive strains of chewing the masticatory muscles grow at an accelerated rate. Ultimately, the muscle grows more than the bone. Dimorphism of the musculature is also evident: the percentage of the head that is muscle is twice as great in males as in females.

This muscle/bone/brain dimorphism contributes to the dimorphism in mass which is evident in 3- to 8-kg animals. The male head is significantly heavier because of the tissues, but as males grow beyond 8 kg, the rest of the body outgrows the head and its components, and the head drops further in relative weight.

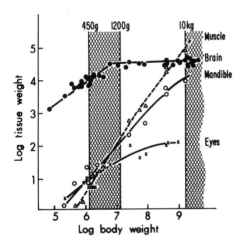

Fig. 6. Growth rates of the tissues of the head (brain, eyes, masticatory muscle, mandible) in relation to weight

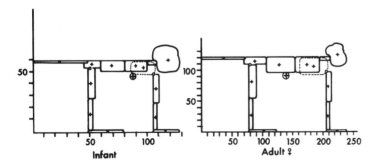

Fig. 7. Positional shift in the center of gravity in infant and adult

   The cross-sectional sample which demonstrates the shift in center of mass is one step removed from the shift in the growing individual. Nevertheless, if the trunk and limbs are drawn to the same linear scale and if the mass centers of the segments are similar, the remaining variable is the relative mass of each segment. The center of gravity (Fig. 7) moves posteriorly about 10% the length of the body. Locomotor capacity and effectiveness at any age or size depend on the position of the center of gravity and the relative strength of the muscle groups.

Discussion

   The increase in weight from 400 g at birth to 8 kg in maturity is not accompanied by uniform growth rates of tissues or segments. The brain and eyes grow to mature size early, muscle groups much later, fatty tissue appears with reproductive maturity. Fowler (1968) pointed out that those components grow apace which are tied by the same functional demands. The macaque counterpart of Fowler's domestic animal sequence is: (1) immediate sensory processing and information storage; (2) emerging nutritional independence; (3) locomotor independence; (4) reproductive maturity; and (5) role dimorphism of males and females.
   The infants, 400 to 800 g and up to 4 months of age, must process both social and visual information. Consequently, the brain and eyes attain nearly mature size shortly after birth. The head is 25% of body weight, the brain 50% of the head. By contrast, the masticatory muscle is less than 1% of the head, the mandible slightly more. Because of physical immaturity, with the body less than 25% muscle, the newborn clings to his mother in order to be carried from place to place. The nuchal extensor muscles, which support the head, and the forearm and calf muscles, which flex the fingers and toes, are disproportionately large. Segmental disproportions reflect this muscular immaturity. As a result of the massive head and the light hindlimbs, the animal is top-heavy (or front-heavy), and the forelimbs do more pulling in quadrupedal walking than the hindlimbs do pushing (Kimura et al. 1979). This is a dependent, rather inefficient locomotor level with no significant sex dimorphism, and only the earliest exploration of physical capacity.
   The juveniles, 1000 g and up to 2 years of age, process food and move about independently. Mandible and masticatory muscles have begun to grow, and the brain and eyes have achieved mature size. Total muscle rises to 40% or 45% of weight, and the juvenile is exceptionally strong. Propulsive muscles of the thigh, hip, back, upper arm, and shoulder grow disproportionately. The heart grows with the musculature and there is increased cardiovascular demand. The center of gravity moves posteriorly so that the hindlimbs push the body more efficiently. These animals explore and exploit their emerging physical skill. When juveniles become sexually mature, bodily dimorphism appears.
   Mason (1979) points out an interesting dichotomy in the external world which parallels one in the developmental sequence for the tissues. The societal world is unpredictable, semierratic, moody. The structural world (the forest canopy), by contrast,

is stable, fixed in space and relatively permanent in time, emotionally neutral. The nervous system has more degrees of freedom because of its neuronal populations and their special sensitivity to early experience. Immaturity extends social experience and thereby promotes behavioral flexibility. By contrast, the locomotor tissues have built-in limitations (e.g., muscle proportions, lever angles, joint surfaces) for the fixed world of trees and terminal branches. The development of locomotor tissues can be delayed, while early development of the brain has adaptive significance.

Adult females, 3- to 8-kg animals, have tissue and segmental maturity. Females may have a higher proportion of body fat developing at this time than males. In any case, adiposity leads to a slight drop in the proportion of muscle to total weight. The head is relatively smaller while the hindlimbs are twice as large as the forelimbs. The center of gravity moves more posteriorly. In this class, the motor repertoire is progressively narrowed and play declines.

In mature males (over 8 kg) and more than 7 years of age, the head is small, the thighs even larger than in mature females. The male brain is absolutely larger than the female brain, and the male masticatory musculature may be twice as large. In free-ranging animals great body size may be a locomotor inconvenience. For example, the largest males of a troop, to avoid slender branches and large gaps they cannot cross because of body weight, make detours which the rest of the troop does not make.

Thus, each stage of life is integrated: the mother's behavior accomodates to her helpless newborn; the juvenile plays and explores both his capacity and his world while his musculature is increasing disproportionately; adipose tissue appears during sexual maturity, perhaps as a buffer to environmental stress. At this latter time, evolution seems to favor adult survival and further reproduction over the newborn who faces long odds before he is ready to reproduce. This dramatic transformation also explains the skill differences which simultaneously must exist within a single troop.

As a matter of fact, the anatomical and skill transformation is so great that, if newborn and adult were not connected by developmental process, I would say that the animals were unrelated. There is a comparable transformation from the stagger of the 1- to 2-year old child to mature walking. Along with changes in proportion of muscle, the distribution of mass, and the location of the center of gravity, the infant must learn the reciprocal discharge of some muscle groups, the suppression of others, stability, and efficient heel-toe-swing action. It takes 5 to 6 years before the pattern "has become almost similar to the adult one" (Okamoto 1973).

The evolutionary lengthening of immaturity in macaques, despite advantages which accompany increasing body size and psychosocial flexibility, exposes the growing animal to other exogenous hazards. The longer the developmental time, the greater the chance that tissue maturation will be deflected by perturbations of the food supply. Early in development, food reduction can affect the brain. Later, reduction affects bone and muscle growth; still later reductions affect the depositon of fat (Widowson 1968, Berg 1968). Whether food reduction simply delays maturation or actually damages critical tissue is debated.

In any event, comparative evidence shows that adipose tissue serves as a buffer against environmental stress. Evolution, it seems, wants to protect reproductively mature animals with internal energy reserves. Thus, stored fat allows Canadian geese to undertake their annual migration, but also permits females to initiate nesting before food is freely available. Clutch size is closely attuned to food supply and reduced food intake may cause the female to stop laying (Raveling 1979). Comparable relations between food supply and human growth and reproduction have been discussed.

Thus, Laudrie (1975) reports periodic increases in sterility (including the cessation of menstruation or amenorrhea) under various economic and political circumstances (famines, siege during wars, extreme price rises in foodstuffs) and uses the term "famine amenorrhea" to characterize the syndrome. Females store almost twice as much fat as males and Frisch (Howell 1979) suggests that a "critical level of fatness" may exist. If a woman falls below this level (approximately 20% of body weight), her fat stores become depleted and reproductive function ceases. In fact, amenorrhea and sterility are related problems faced by endurance athletes (e.g., marathoners) in heavy training.

Malina and Zavaleta (1980) report a broad increase in height and weight among the children of most ethnic groups throughout the industrialized world during the period from 1930 to 1970. However, Mexican-Americans have not grown as much as other groups which indicates that as a population they receive food of relatively lower nutritional value.

To conclude, I want to emphasize the internal consistency of these anatomical techniques which make living weight the baseline and deal with the common currency of tissue composition (for the nutritionist and physiologist) and segmental masses (for the functional anatomist). Ross, Martin, Drinkwater and Clarys (pers. comm.) have begun comparable studies of adult humans at the Free University of Brussels.

Finally, within the limits of the methods discussed, my data and functional interpretations are probably correct. However, their importance really depends upon the questions which this research raises. What are the adaptive values of attenuated anatomical and locomotor maturity? What are the effects of undernutrition for each developmental stage? How is energy intake at each stage of life balanced with metabolic demands for maintenance, for growth, and for activity? As the troop moves along, what are the special locomotor problems of each skill class? Do macaque females possess more body fat than males and does this additional fat correlate with the energy demands of gestation and labor?

Acknowledgments. Several friends and colleagues have discussed this work: Charles Kowalski, Charles Menzel, Neal Cross, Bill Ross, Alan Martin. Janina Ely polished the manuscript and Wy Holden typed the final copy; Jeannette Cissman did the final figures. This work, Publication no. 1170 of the Oregon Regional Primate Research Center, was supported by NIH no. RR-00163.

References

Berg RT (1968) Genetic and environmental influences on growth in beef cattle. In: Lodge GA, Lamming GE (eds) Growth and development of mammals. Plenum Press, New York,London, pp 429–450

Butterfield RM, Johnson ER (1968) The effect of growth rate of muscle in cattle on conformation as influenced by muscle-weight distribution. In: Lodge GA, Lamming GE (eds) Growth and development of mammals. Plenum Press, New York London, pp 212–223

Fowler VR (1968) Body development and some problems of its evaluation. In: Lodge GA, Lamming GE (eds) Growth and development of mammals. Plenum Press, New York London, pp 195–211

Grand TI (1977a) Body weight: its relation to tissue composition, segment distribution, and motor function. I. Interspecific comparisons. Am J Phys Anthropol 47:211–240

Grand TI (1977b) Body weight: its relation to tissue composition, segment distribution, and motor function. Am J Phys Anthropol 47:241–248

Grand TI (1978) Adaptations of tissues and limb segments to facilitate moving and feeding in arboreal folivores. In: Montgomery GG (ed) The ecology of arboreal folivores. Smithsonian Inst Press, Washington DC, pp 231–241

Howell N (1979) Demography of the Dobe: Kung. Academic Press, London New York, 389 pp

Kimura T, Okada M, Ishida M (1979) Kinesiological characteristics of primate walking and its significance in human walking. In: Morbeck ME, Preuschoft H, Gomberg N (eds) Environment, behavior, and morphology: Dynamic interactions in primates. G Fischer, New York Stuttgart, pp 297–312

Kowalski CJ, Guire KE (1974) Longitudinal data analysis. Growth 38:131–159

Laudrie E LeRoy (1975) Famine amenorrhea (seventeenth-twentieth centuries). In: Forster R, Ranum O (eds) Biology of man in history. Johns Hopkins Press, Baltimore, pp 163–178

Malina R, Zavaleta AN (1980) Secular trend in the stature and weight of Mexican-American children in Texas between 1930 and 1970. Am J Phys Anthropol 52:453–462

Mason W (1979) Ontogeny of social behavior. In: Marler P, Vandenberg JG (eds) Handbook of behavioral neurobiology, vol I. Social behavior and communication. Plenum Press, New York London, pp 1–28

Miller DI, Nelson RC (1973) Biomechanics of sport. A research approach. Lea and Febiger, Philadelphia, 265 pp

O'Connor RJ (1975) The influence of brood size upon metabolic rate and body temperature in nestling Blue Tits, *Parus caeruleus,* and House Sparrows, *Passer domesticus.* J Zool 175:391–403

O'Connor RJ (1977) Growth strategies in nestling passerines. Living Bird XVI:209–238

Okamoto T (1973) Electromyographic study of the learning process of walking in 1- and 2-year old infants. In Cerquiglini S, Venerando A, Wartenweiler J (eds) Medicine and sport, vol 8: Biomechanics III, Karger, Basel, pp 328–333

Palmer CE (1944) Studies of the center of gravity in the human body. Child Dev 15:99–180

Raveling DG (1979 The annual cycle of body composition of Canada geese with special reference to control of reproduction. Auk 96:234–252

Ricklefs RE (1968) Patterns of growth in birds. Ibis 110:419–451

Ricklefs RE (1973) Patterns of growth in birds. II. Growth rate and mode of development. Ibis 115:177–201

Tanner JM (1969) Growth at adolescence. Blackwell Scientific Publ, Oxford, 326 pp

Widdowson EM (1968) The effect of growth retardation on postnatal development. In: Lodge GA, Lamming GE (eds) Growth and development of mammals. Plenum Press, New York London, pp 224–233

Morphological and Ecological Characters in Sympatric Populations of Macaca in the Dawna Range

A.A. Eudey [1]

Within the order Primates many instances of sympatry of congeneric species have been recorded. Most efforts to analyze these occurrences have been influenced by the principles of competitive exclusion (cf. Hardin 1960) or character divergence/character displacement (Darwin 1958, pp 111–125, Brown and Wilson 1956) and, as a consequence, have concentrated on detailing differential utilization of environmental resources by the sympatric populations and/or morphological differences between them. With few exceptions (cf. Sussman 1979), the evolutionary events, although speculative, that may have led to such sympatry or overlap in ranges, are given minimal or no consideration.

In 1973 I initiated field studies on macaques (*Macaca* spp.) in a known area of sympatry (Fooden 1971) within the boundaries of Huay Kha Khaeng Game Sanctuary between 15° and 16° N latitude in the Dawna Range in west-central Thailand (Fig. 1). The Dawna Range is characterized by monsoon deciduous forest, frequently with much bamboo, and dry evergreen forest, patchily distributed from lowland areas to mountainous regions of limestone formation reaching elevations of 1200 to 1500 m. A detailed description of the area appears in Eudey (1979).

Over the 7-year period 1973–1979, I completed five field sessions, ranging in duration from 6-month to approximately 3-week intervals (Table 1). During the two longest field sessions, residence in the game sanctuary was discontinuous. All months have been sampled with the exception of June and September, both of which fall within the 6-month monsoon season in the area.

A total of 154 "search days" (Table 1) was spent attempting to contact macaques in representative habitats within an area of, at a minimum, 20 km^2, much of which consists of precipitous terrain. Only 52 contacts were made on 39 search days with groups of macaques ranging in size from less than 10 to at least 50, or with one or two individual monkeys. Some of these contacts were multiple encounters with the same group or individual(s). Duration of contacts lasted from 30 s to 315 min. The short duration of contacts may be attributed to the fact that no groups were habituated, while the limited number of contacts appears to be a consequence of the low densities of all populations.

Approximately 73% of contacts (38 of 52) were made in the vicinity of the Khao Nang Rum Research Station (15°29′ N latitude, 99°17.5′ E longitude) in mountainous areas between 300 and 800 m where deciduous, bamboo, and dry evergreen forests

1 Department of Anthropology, University of Nevada, Reno, Nevada 89557, USA

Table 1. Summary of searches for and contacts with macaques (*Macaca* spp.) in Huay Kha Khaeng Game Sanctuary during the 1973–1974, 1975, 1977, 1978, and 1979 field sessions

	Search days	Contact days	Total number of contacts	Search hours (sh)	Contact hours (ch)	% ch/sh	Range of contacts in min	Average contact in min
Oct 1973 to Mar 1974	65	12	14	300 h	8 h 28 min	2.8	1.0 to 315	36
Jan 1975 to May 1975	46	6	8	195 h	6 h 44 min	3.5	1.0 to 111	51
13 Jul to 21 Aug 1977	19	7	9	88 h	4 h 55 min	5.6	2.0 to 136	33
14 Jul to 30 Jul 1978	10	5	6	44 h 40 min	2 h 27 min	5.5	1.0 to 120	24.5
15 Jul to 5 Aug 1979	14	9	15	69 h 22 min	9 h 27 min	13.6	0.50 to 84	37.8
Totals	154	39	52	697 h 2 min	32 h 1 min	4.6	0.50 to 315	36.9

Table 2. Number of times search areas in Huay Kha Khaeng Game Sanctuary were traversed during the 1973–1974, 1975, 1977, 1978, and 1979 field sessions. Search areas are indicated on Fig. 2. Number of contacts with macaques in each area is indicated in parentheses

	A	B	C	D	E	F	G	H	I	J	K	Total contacts
1973–1974	4 (1)	2 (2)	2 (1)	2 0	20 (1)	4 (1)	7 (3)	11 (4)	9 (1)	4 0		(14)
1975	15 (1)	1 0	3 (1)	3 0	18 (1)	1 0	10 (5)					(8)
1977	2 0		1 0	2 0	2 0	8 (6)	5 (2)				2 (1)	(9)
1978			1 0	2 0	1 0	2 0	5 (6)					(6)
1979				1 0		3 (2)	7 (13)				1 0	(15)
Total times surveyed	21 (2)	3 (2)	7 (2)	10 0	41 (2)	18 (9)	34 (29)	11 (4)	9 (1)	4 0	3 (1)	(52)
Percentage of contacts	3.85	3.85	3.85	0	3.85	17.3	55.8	7.7	1.9	0	1.9	100

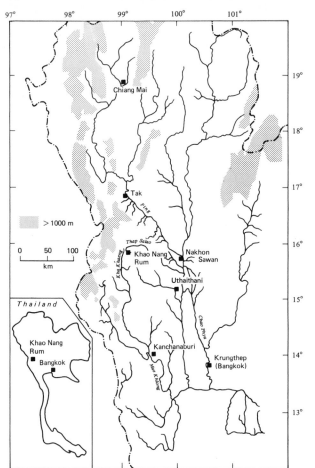

Fig. 1. Location of Huay Kha Khaeng Game Sanctuary, as indicated by the research station at Khao Nang Rum, in the Dawna Range in west-central Thailand

intergrade (Fig. 2, Table 2). These contacts (Table 3) were predominantly with *Macaca assamensis* and *M. nemestrina leonina* and rarely with *M. mulatta* and *M. arctoides,* the latter seemingly occurring at higher elevations. Both mixed species and apparent hybrid groups also were recorded in these areas. *M. fascicularis* were recorded only in lowland areas, immediately adjacent to or at a distance of up to 2000 m from the one major river [*huay* (= river) Thap Salao] in the search area, but within the reported range of *M. assamensis* and *M. nemestrina leonina* and, probably, the range of *M. mulatta.*

Macaca assamensis and *M. nemestrina leonina* converge in some morphological characters: red dorsal pelage occurs in both populations although it is almost universal and darker in *M. assamensis;* body size in *M. assamensis* is large and robust while medium to large in *M. nemestrina leonina;* both populations are short-tailed although *M. assamensis* exhibits variation in tail length. The other three species are more divergent in some morphological characters although three groups recorded may be short-tailed *M. fascicularis* or *fascicularis-mulatta* hybrids. The few *M. arctoides* recorded,

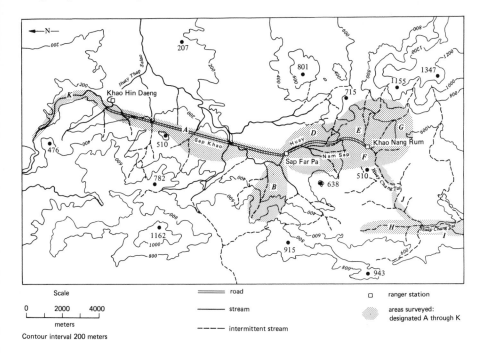

Fig. 2. Northeastern area of Huay Kha Khaeng Game Sanctuary, with areas searched for macaques from 1973–1974 to 1979 designated *A* through *K*. The Khao Nang Rum Research Station is located at $15°29'$ N latitude, $99°17.5'$ E longitude

presumably a solitary male and a small group of two or three males, are black with black faces. Populations are described in detail in Eudey (1979).

All macaques appear to exploit large spatio-temporal ranges that are altitudinally graded, encompassing both lower elevations and mountainous regions, and which permit them to sample different habitats for seasonally available fruits, the principal dietary component during all months surveyed (Eudey 1979, 1980). Similar far-ranging patterns have been recorded for other forest macaques, e.g., *Macaca mulatta* in north India (Lindburg 1977) and *M. silenus* in the Western Ghats of south India (Green and Minkowski 1977). However, in comparison with the other species of macaques in the Dawna Range, *M. fascicularis* may be less far ranging and restricted to riverine areas. Such a restriction to riverine habitat has been reported for populations of *M. fascicularis* in both Borneo (Rodman 1973, 1977, Fittinghoff and Lindburg 1980, Wheatley 1980) and Sumatra (Rijksen 1978, Crockett and Wilson 1980).

No gross differences in the pattern of habitat utilization are, to date, apparent among the different macaques, i.e., they do not exclude each other from environmental resources, but *Macaca nemestrina leonina* forages in subgroups and *M. assamensis* forages in either phalanx or file formations in apparent response to dispersal of fruits. Subgroups, as recorded in *M. nemestrina leonina*, also have been reported for populations of *M. nemestrina nemestrina* (Rijksen 1978, Crockett and Wilson 1980), and *M. silenus* (Sugiyama 1968). Foraging in subgroups may be under phylogenetic

Table 3. Summary of number of contacts with different species of macaques in Huay Kha Khaeng Game Sanctuary from 1973 to 1979 by search areas designated A through K on Fig. 2

Species	A	B	C	D	E	F	G	H	I	J	K	Total contacts
M. arctoides		1					1					2
M. assamensis [a]					1	3	10		1			15
M. fascicularis	2										1	3
M. mulatta						1						1
M. nemestrina leonina [a]							6					6
M. mulatta or short-tailed *M. fascicularis* or *mulatta-fascicularis* hybrids			1		1			1				3
Other groups of mixed species or hybrid composition [b]						1	4					5
Unknown [c]		1	1			4	8	3				17
Total contacts	2	2	2	0	2	9	29	4	1	0	1	52

a   Some groups identified as *M. assamensis* or *M. nemestrina leonina* may have contained monkeys of the other species and/or hybrids

b   Most groups exhibiting mixed morphology and/or more than one species contained recognizable members of *M. assamensis*

c   Most unidentifiable monkeys were one or two individuals rather than groups

constraints as *M. silenus* and *M. nemestrina* appear to be related and are assigned to the same species group on the basis of morphological characters (Fooden 1980, Delson 1980).

The dispersal and, probably, the differentiation of Asian macaques appear to be associated with glacial phenomena of the Pleistocene epoch, especially climatic deterioration and associated deforestation occurring at latitudes south of the continental ice sheets. Elsewhere (Eudey 1979, 1980) I have proposed on the basis of present distributions that the Dawna Range, along with the Khasi Hills in Assam and the Annamitic Cordillera in Viet Nam, constituted a major forest refugium for macaques (and other mammals) during such periods of deterioration.

Macaca nemestrina leonina probably first invaded the Dawna Range, followed by *M. assamensis* and *M. arctoides,* which may be related to each other (Delson 1980). Competition between these latter two species may have contributed to the limited representation of *M. arctoides* in the study area and elsewhere. The red coloration (erythrism) of *M. nemestrina leonina* and *M. assamensis* may contribute to the crypticity of these species in deciduous forest during dry seasons, and convergence in this morphological character may reflect the antiquity and extent of the sympatry of these two populations in the Dawna Range. *M. mulatta* and *M. fascicularis*, which are assigned to the same species group (Fooden 1980, Delson 1980), appear to have entered the area most recently. *M. fascicularis* probably reached the Dawna Range by dispersing eastward and northward through forests fringing major rivers and streams during the last glacial (about 20,000 years ago) after having undergone significant differentiation in Sundaland (Eudey 1979, 1980). The preference for, or restriction to, riverine habitats in populations of *M. fascicularis* appears to date back to at least the late Pleistocene dispersal of the species and may be under phylogenetic constraints. The marked seasonality of the Dawna Range appears to test these constraints and necessitates similar ecological adaptations by all macaques which have sought refuge in the area.

Convergence or divergence of morphological and ecological characters in an area of sympatry such as the Dawna Range may best be explained in terms of the interplay of phylogenetic constraints and the unique series of evolutionary events that led to the observed instance of sympatry.

Acknowledgments. Fieldwork was supported by the Wenner-Gren Foundation and Sigma Xi in 1973–1974 and the National Science Foundation, the National Institute of Mental Health, and Sigma Xi in 1975. My participation in the development of a research station at Khao Nang Rum was made possible by awards from the New York Zoological Society in 1977 and 1979 and the Fauna Preservation Society in 1977. I wish to thank the National Research Council and the Wildlife Conservation Division, Royal Forest Department, for sponsoring my research in Thailand.

References

Brown WL, Wilson EO (1956) Character displacement. Syst Zool 5:49–64

Crockett CM, Wilson WL (1980) The ecological separation of *Macaca nemestrina* and *M. fascicularis* in Sumatra. In: Lindburg DG (ed) The macaques: studies in ecology, behavior and evolution. Van Nostrand Reinhold, New York, pp 148–181

Darwin C (1958) The origin of species. Mentor Book, New Am Library, New York

Delson E (1980) Fossil macaques, phyletic relationships and a scenario of deployment. In: Lindburg DG (ed) The macaques: studies in ecology, behavior and evolution. Van Nostrand Reinhold, New York, pp 10–30

Eudey AA (1979) Differentiation and dispersal of macaques (*Macaca* spp.) in Asia. Doctoral dissertation. Dep Anthropol, Univ California, Davis

Eudey AA (1980) Pleistocene glacial phenomena and the evolution of Asian macaques. In: Lindburg DG (ed) The macaques: studies in ecology, behavior and evolution. Van Nostrand Reinhold, New York, pp 52–83

Fittinghoff NA, Lindburg DG (1980) Riverine refuging in east Bornean *Macaca fascicularis*. In: Lindburg DG (ed) The macaques: studies in ecology, behavior and evolution. Van Nostrand Reinhold, New York, pp 182–214

Fooden J (1971) Report on primates collected in western Thailand January–April, 1967. Fieldiana Zool 59:1–62

Fooden J (1980) Classification and distribution of living macaques *(Macaca Lacépède* 1799). In: Lindburg DG (ed) The macaques: studies in ecology, behavior and evolution. Van Nostrand Reinhold, New York, pp 1–9

Green S, Minkowski K (1977) The lion-tailed monkey and its south Indian rain forest habitat. In: Prince Rainier of Monaco HSH, Bourne GH (eds) Primate conservation. Academic Press, London New York, pp 289–337

Hardin G (1960) The competitive exclusion principle. Science 131:1291–1297

Lindburg DG (1977) Feeding behaviour and diet of rhesus monkeys *(Macaca mulatta)* in a Siwalik forest in north India. In: Clutton-Brock TH (ed) Primate ecology: studies of feeding and ranging behaviour in lemurs, monkeys, and apes. Academic Press, London New York, pp 223–249

Rijksen HD (1978) A field study on Sumatran orang utans *(Pongo pygmaeus abelli* Lesson 1827); ecology, behaviour and conservation. H Veenman and Zonen BV, Wageningen

Rodman PS (1973) Synecology of Bornean primates. I. A test for interspecific interactions in spatial distribution of five species. Am J Phys Anthropol 38:655–659

Rodman PS (1977) Food distribution and terrestrial locomotion of crab-eating and pig-tailed macaques in the wild. Am J Phys Anthropol 47:157

Sugiyama Y (1968) The ecology of the lion-tailed macaque [*Macaca silenus* (Linnaeus)] – a pilot study. J Bombay Nat Hist Soc 65:283–292

Sussman RW (ed) (1979) Primate ecology: problem-oriented field studies. John Wiley and Sons, New York

Wheatley BP (1980) Feeding and ranging of east Bornean *Macaca fascicularis.* In: Lindburg DG (ed) The macaques: studies in ecology, behavior and evolution. Van Nostrand Reinhold, New York, pp 215–246

Specialization of Primate Foot Reflected in Quantitative Analysis of Arthrodial Joints of Anterior Tarsals

D.K. Messmann [1]

Tarsal joints are mainly responsible for inversion and eversion of the foot. They also are responsible for the changes in configuration as the foot is applied to the ground or takes hold of an object as in climbing (Gray 1910). Consequently, consistent use of the tarsal in a locomotion pattern may be reflected in some aspects of the articular surfaces.

Locomotion was one of the prime considerations in selection of genera for a quantitative analysis of tarsal bones. Arboreal and/or terrestrial representatives of the principal locomotor categories of living primates — quadrupedalism, brachiation, and bipedalism — were selected. Initial research has been limited to extant forms, but the research objective is to obtain interpretative data for application to fossil forms, especially hominoids. As a result, there is an emphasis on apes and a slighting of monkeys, especially New World monkeys. Analysis of the following genera are emphasized in this report: *Ateles, Cebus, Colobus, Gorilla, Homo, Hylobates, Pan, Papio, Pongo,* and *Presbytis.*

A minimum sample size of 40 for each genus of nonhuman primates was sought in the American Museum of Natural History (New York), Field Museum of Natural History (Chicago), and the National Museum of Natural History (Washington D.C.). This goal was not reached for an arboreal quadrupedal Old World monkey using *Colobus.* Therefore, *Presbytis* was included in the study to ensure an adequate sample of arboreal Old World monkeys. Inclusion of *Presbytis* also permitted observation for similarity. For further insight into relative similarties, small samples (2–10) of *Mandrillus, Symphalangus,* and *Theropithecus* were also used.

Hominid specimens measured included a population of American Woodland Indians (28) from the Department of Anthropology and modern whites (38) from the Department of Physiology, University of Illinois. Negroid measurements (12) were obtained from specimens of the American Museum of Natural History. Total of hominid specimens was 78.

Subjects for study were not limited to specimens in their prime, but included immature, aged, infirm, and diseased. This was done for two reasons: (1) because the research goal is application to the fossil record and animals dying in their prime are not most common, and (2) because an appreciation of the extent of bone plasticity is being sought for future research. The latter reason has also prompted use of zoo specimens, as well as those collected in the wild, for some aspects of the research.

---

1 Department of Anthropology, University of Illinois, Urbana, IL 61801, USA

The majority of specimens are adult. They include left and right tarsal bones from males and females selected randomly and not in equal numbers. Although all the tarsals were measured, this report will be limited to variables of the first cuneiform and cuboid. Height, length, and width of a bone, or a feature on a bone, were established relative to anatomical position.

Five examples have been selected to illustrate the range and potential of research on individual tarsal bones. These include: (1) examination for absence and presence of traits related to mobility potential, (2) analysis of relative proportions of a trait in terms of mobility or stability, and (3) consideration of articular surfaces as indicators of stress direction and movement range.

One of the most obvious examples of the informative value of an articular surface occurs in the first cuneiform. Earlier research (Lewis 1972) has suggested that the grade reached by New World monkeys and *Dryopithecus* in the hallucial tarsometatarsal joint has been further modified in cercopithecoid and hominoid evolution by the grasping ability of the hallux. If this is the case, a noticeable difference may be evident when comparing the first cuneiforms of New and Old World monkeys. A more marked difference should be obvious when extending the comparison to apes with their specialized grasping foot and to humans with their specialized foot for bipedalism.

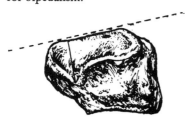

Fig. 1. Medial view of the right first cuneiform of *Pan troglodytes,* Chicago Field Museum Natural History Specimen 51319 with the plane of the axis and the medial overlap marked on the hallucial metatarsal articular surface. Drawing by Marylin Weiss. Actual size

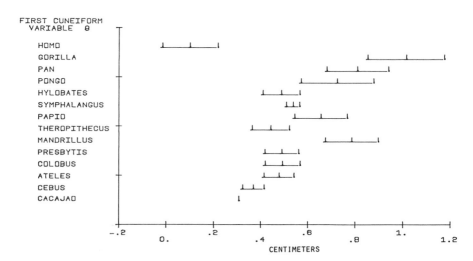

Fig. 2. Medial overlap in centimeters of the hallucial tarsometatarsal articular surface of the first cuneiform. *Central marker on each line* represents mean; *end markers* represent one standard deviation

Measurement of the variable of medial extension beyond the plane of the axis of the articular surface of the first cuneiform with the hallucial metatarsal (Fig. 1) has been labeled medial overlap. This measurement (Fig. 2) reflects abduction mobility of the joint. Lateral overlap (Fig. 3), similarly measured laterally, relates to adduction mobility of the joint. Maximum length of the first cuneiform was measured after Hrdlička (1920) (Fig. 4) for comparison to length of head and body (Fig. 5) to

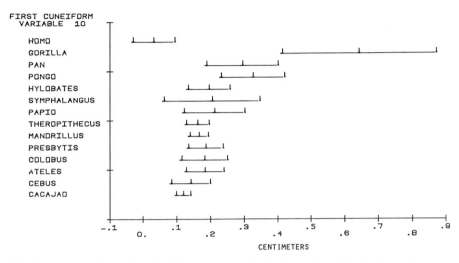

Fig. 3. Lateral overlap of the hallucial tarsometatarsal articular surface of the first cuneiform in centimeters. *Central marker on each line* represents mean; *end markers* represent one standard deviation

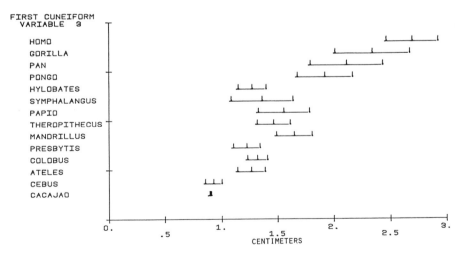

Fig. 4. Maximum length in centimeters of the first cuneiform obtained by applying the fixed branch of the calipers to the most prominent part of the inferior surface of the bone and bringing the other branch into opposition (Hrdlička 1920). *Central marker on each line* represents the mean; *end markers* represent one standard deviation

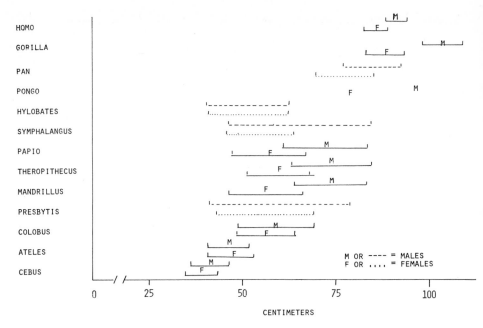

Fig. 5. Head and body length. *Central letter on line* represents mean; *end markers* represent one standard deviation. *Broken* or *dotted lines* represent range. (Data from Clauser et al. 1972, Hertzberg and Daniels 1954, Hill 1960, 1962, 1970, Napier and Napier 1967, Roonwal and Mohnot 1977, Willoughby 1978)

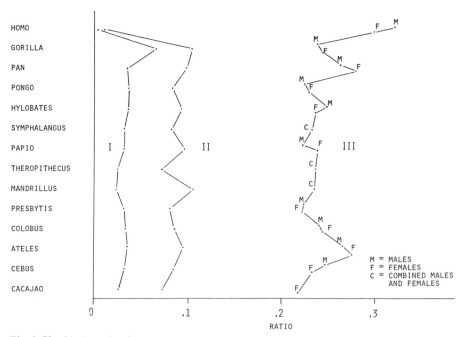

Fig. 6. *Plot I* is the ratio of lateral overlap to head and body length. *Plot II* is the ratio of medial overlap to head and body length. *Plot III* is the ratio of cuneiform length to head and body length. Relative to head and body length, *Homo* has the longest foot, but the shortest medial and lateral overlap

control for the different body sizes of the genera being studied. The three variables:
(1) cuneiform length, (2) medial overlap, and (3) lateral overlap were plotted as ratios
to head and body length (Fig. 6). Results are most striking for *Homo;* proportionally
this genus has the longest foot and the least medial and lateral overlap. The loss of
convexity in *Homo* for this joint as illustrated by medial overlap (Fig. 5) and lateral
overlap (Fig. 6) is obvious in a cursory comparison of the first cuneiform of *Homo*
to any ape. In *Homo* the hallucial tarsometatarsal articular surface has become essentially flat; mobility for grasping has been sacrificed for stability and weight bearing.
Varying degrees of this trait exhibited in the fossil record should indicate evolutionary trends in pongid and hominid adaptations.

A more subtle example of extracting information from a single bone can be
demonstrated using the smallest articular surface of the first cuneiform, the articular
surface with the second metatarsal. The ratio of the length of the second metatarsal
articular surface to the maximum length of the bone, or the percent of the bone
length that is articular surface for this joint, has been plotted (Fig. 7). The smallest
mean percents are *Theropithecus* (21%) and *Homo* (25%).

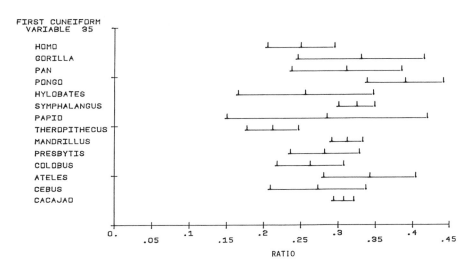

Fig. 7. Plot of the ratio of the length of the first cuneiform-second metatarsal articular surface to
the maximum length of the first cuneiform. *Central marker on line* represents mean; *end markers*
represent one standard deviation

The ratio of the height of the second metatarsal articular surface to the maximum
height of the bone has also been plotted (Fig. 8). Smallest mean percents are *Homo*
(20%), *Mandrillus* (21%), and *Cebus* (21%). A sample of 50 *Cebus* specimens has been
used for this study. The height ratio of the second metatarsal articular surface is only
one example of findings difficult to interpret for *Cebus,* but information from
specialists and further research may resolve the problem.

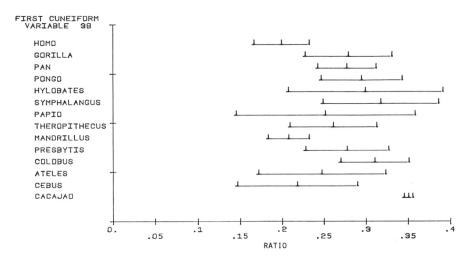

Fig. 8. Plot of the ratio of the height of the first cuneiform-second metatarsal articular surface to the maximum height of the bone. *Central marker on line* represents mean; *end markers* represent one standard deviation

The relatively small area of second metatarsal articulation in the genera mentioned above means a decrease in the amount of gliding and sliding possible between these two bones, which again suggests emphasis on stability over mobility. Great apes have the greatest potential for mobility for the second metatarsal at this joint. If both height and width of the second metatarsal articular surface are considered, terrestrial Old World monkeys have less articular surface area and decreased mobility compared with the arboreal Old World monkey, New World monkeys (except *Cebus*), and lesser apes. *Homo* has the smallest articular surface proportionally and the greatest decrease in mobility.

The numerous small jointed bones in the foot give it elasticity (Lockhart et al. 1959). Although the articular surfaces are small, monkeys have a greater number of joints than apes and man on the second and third cuneiforms. One example is the percent of specimens in each genus that has an anterior articular surface between the third cuneiform and the cuboid (Table 1). Except for *Cebus,* all monkey specimens had an anterior articular surface between the third cuneiform and the cuboid, as well as a posterior one. In the great apes, presences of this articular surface ranged from 86.0% to 90.6%, but in *Homo* presence is rare, 1.3%. Numerous articular surfaces in the anterior tarsal region maximize mobility for grasping; conversely, lack of articulation means loss of movement.

All of the tarsal joints, except the one between the talus and navicular are classified as arthrodial joints permitting gliding movement only. However, gliding movement in an arthrodial joint need not be confined to a plane surface, but may exist between any two contiguous surfaces of whatever form (Gray 1910). As seen in the hallucial tarsometatarsal joint of the first cuneiform, the flatter articular surface in *Homo* has decreased mobility compared with the more convex articular surfaces of

Table 1. Presence of an anterior third cuneiform articular surface on the cuboid

Genera	Percentage of specimens with an anterior articular surface
Homo	1.3%
Gorilla	86.7%
Pan	90.6%
Pongo	86.0%
Hylobates	100.0%
Papio	100.0%
Theropithecus	100.0%
Mandrillus	100.0%
Presbytis	100.0%
Colobus	100.0%
Ateles	100.0%
Cebus	87.8%

apes and monkeys. Similarily, a shallow articular surface of the fourth metatarsal with the cuboid decreases mobility, whereas a deeper, curved surface increases the range of gliding movement possible. Depth measurements of the cuboid-fourth metatarsal articular surface have been measured and plotted (Fig. 9). If relative cuboid and body lengths are considered (Fig. 10) *Homo* and the terrestrial monkeys have shallow articular surfaces, whereas lesser apes have relatively deep ones.

Gliding movement in an arthrodial joint is limited by the ligaments which enclose the articulation or osseous processes surrounding the articulation (Gray 1910). The posterior plantar projection on the medial side of the cuboid confines movement at

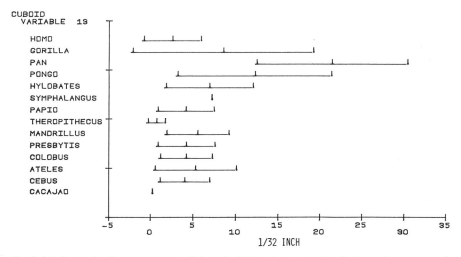

Fig. 9. Maximum depth measurement of the cuboid-fourth metatarsal articular surface expressed in 1/32 of an inch. *Central marker on line* represents mean; *end markers* represent one standard deviation

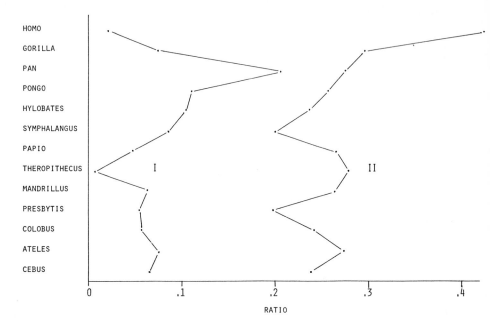

Fig. 10. *Plot I* is the ratio of the depth measurement of the cuboid-fourth metatarsal articular surface to head and body and body length. *Plot II* is the ratio of cuboid bone length to head and body length

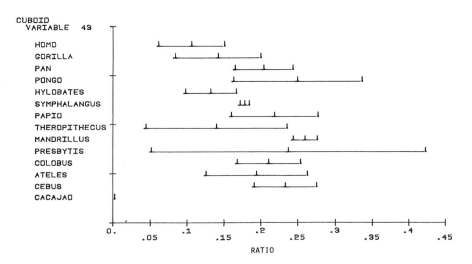

Fig. 11. Plot of the ratio of the posterior plantar projection of the cuboid to the maximum length of the cuboid. *Central marker* represents mean; *end markers* represent one standard deviation

the articular surface between the cuboid and the calcaneus. When the ratio of this projection to maximum length of the bone is plotted (Fig. 11), the projection is the smallest percent of the bone length in *Homo* (10%) suggesting less confinement of movement. Genera with bipedal tendencies generally have projections less than 20%, whereas strictly quadrupedal genera have projections over 20% of the bone length. The relatively shorter projection in *Homo* permits more mobility at this joint which is part of the transverse arch. Increased mobility at the transverse arch joined with decreased mobility and greater stability of the cuboid-fourth metatarsal joint is a commitment to upright bipedalism. Occasional bipeds maintain greater mobility in the anterior tarsometatarsal articulations for grasping.

Although not always obvious to visual inspection, bone configurations and articular surface areas exhibit trends in spite of genetic variation and bone plasticity. Articular surfaces on the tarsal bones appear in a bewildering array of shapes, but their relative proportions and contours display interesting trends that correlate with known locomotion patterns. Articular surfaces give insight into evolutionary specialization and may allow interpretation of fossil forms, both those antecedent to modern genera and those whose specializations may not be represented among living primates.

Acknowledgment. Research was made possible by an American Association of University Women Fellowship

References

Clauser CE, Tucker PE, Reardon JA, McConvill JT, Churchill E, Lauback LL (1972) Anthropometry of Air Force Women AMRL-TR-70-5. US Government Printing Office, 1157 pp
Gray H (1910) Anatomy, descriptive and applied, 18th edn. Lea and Febiger, Philadelphia, 1496 pp
Hertzberg HT, Daniels GS (1954) Anthropometry of flying personnel 1950. WADC Tech Rep 52-321. McGregor and Werner Midwest Corp, Dayton, Ohio, 134 pp
Hill WC (1960) Primates, comparative anatomy and taxonomy, vol IV. Cebidae, part A. Wiley Interscience Publ Inc, New York, 523 pp
Hill WC (1962) Primates, comparative anatomy and taxonomy, vol V. Cebidae, part B. Wiley Interscience Publ Inc, New York, 537 pp
Hill WC (1970) Primates, comparative anatomy and taxonomy, vol VIII. Cynopithecinae. Wiley Interscience Publ Inc, New York, 680 pp
Hrdlička A (1920) Anthropometry. Wistar Institute of Anatomy and Biology, Philadelphia, 163 pp
Lewis OJ (1972) The evolution of the hallucial tarsometatarsal joint in the Anthropoidea. Am J Phys Anthropol 37 (1):13–34
Lockhart RD, Hamilton GF, Fyfe FW (1959) Anatomy of the human body. JB Lippincott Co, Philadelphia, 697 pp
Napier J, Napier PA (1967) A handbook of living primates. Academic Press, London New York, 456 pp
Roonwal ML, Mohnot SM (1977) Primates of South Asia. Harvard University Press, Harvard, 421 pp
Willoughby DP (1978) All about gorillas. AA Barnes and Co, New York, 264 pp

Morphology of Some of the Lower Limb Muscles in Primates

A. Prejzner-Morawska and M. Urbanowicz [1]

The purpose of our research was observation of the morphology and the morphological variation of some of the lower limb muscles in primates.

The study material consisted of the extremities (usually both) of 60 primates, including 10 Prosimiae, 15 platyrrhine monkeys, 22 catarrhine monkeys and 13 anthropoid apes. The material of Prosimiae consisted of: *Nycticebus coucang* 1, *Loris gracilis* 1, *Lemur varius* 4, *Lemur bruneus* 3, *Lemur macaco* 1. The platyrrhines included: *Callithrix penicillata* 2, *Cebus capucinus* 6, *Leontocebus rosalia* 2, *Saimiri sciureus* 1, *Ateles ater* 2, *Ateles geoffroyi* 1, *Alouatta* 1. The material of the lower catarrhine monkeys consisted of: *Macaca mulatta* 13, *Cercopithecus aethiops* 5, *Presbytis entellus* 4. The anthropoid apes were: *Symphalangus syndactylus* 2, *Hylobates syndactylus* 2, *Pongo pygmaeus* 1, *Pan troglodytes* 8.

The following muscles were studied: (1) biceps femoris m., (2) tenuissimus m., (3) semitendinosus m., (4) triceps surae m. Of the muscles listed above the semitendinous m. was examined on the smaller number of monkeys as mentioned hereafter.

After incising and dissecting the skin and exposing the muscle, the fascia was removed and the insertion and shape of the muscle were noted. Sketches were made, and interesting varieties were photographed. Several measurements of the muscles were made and the corresponding indexes were calculated (Loth 1931). On the basis of these indexes schemes showing the morphological features of the muscles were drawn.

The Biceps Femoris Muscle

In the majority of primates the biceps femoris muscle had only one head — long head, i.e., ischiocrural lateral muscle. The so-called short head was observed only in some of the platyrrhine monkeys *(Ateles, Alouatta)* and in anthropoid apes (Prejzner-Morawska and Urbanowicz 1971). In lemurs the biceps femoris was represented by a long head stretching from the ischial tuberosity to the tibia. The muscle was slender spreading out fanwise near its terminal insertion. The tibial insertion had the character of a thin aponeurosis passing into the crural fascia (from the level of the knee joint to the middle of the tibia, or lower).

---

1 Department of Normal Anatomy, Institute of Medical Biology, Medical Academy, Gdańsk, Poland

In platyrrhine monkeys (except *Ateles* and *Alouatta*) the ischiocrural lateral muscle was accompanied by the tenuissimus muscle, which is considered to be the homolog of the short head of biceps femoris.

In *Ateles* and *Alouatta* the biceps was similar, possessing both heads (Fig. 1A). The long head in both species began on the ischial tuberosity and ended on the lateral condyle of the tibia. The short head stretched from the linea aspera of the femur, in *Ateles* mainly to the crural fascia, and in *Alouatta* to the lateral condyle of the tibia. At the terminal insertion, in *Ateles* the fibers of both heads intermingled very little and in *Alouatta* fusion between the two heads was considerable (Fig. 1B). In this respect, *Alouatta* differed from all the other primates, including anthropoids, resembling most the relations in Man.

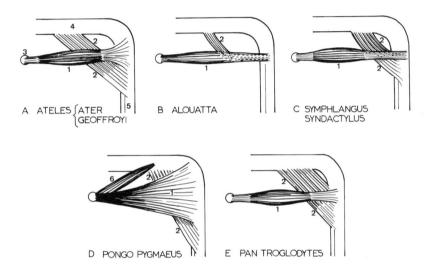

Fig. 1. A–E. Shape and relation of both heads of the biceps femoris muscle in the studied monkeys. Lateral view on the right limb. Scheme: *1*, long head; *2*, short head; *3*, ischial tuberosity; *4*, femur; *5*, ossa cruris; *6*, tuberofemoral bundle of Fick

In the lower catarrhine monkeys the muscle was similar to that in lemurs and most platyrrhines. Only the terminal fascial insertion reached a little higher, being attached not only to the upper part of the tibia, but also to the lower end of the femoral fascia.

Among the anthropoid apes (possessing two heads of the biceps femoris muscle) the degree of fusion between the two heads differed distinctly. The most primitive relations were observed in the orangutan, namely broad insertion of the long head, low terminal insertion of the short head on the tibia, and very slight fusion of the heads. The upper part of the long head separates as the so-called tuberofemoral bundle of Fick (Fig. 1D).

In *Symphalangus*, on the other hand, the relations resembled most those in Man (Fig. 1C). The area of insertion of both heads was maximally reduced, and fusion of both heads the greatest, approaching the degree of fusion observed in Man. In the

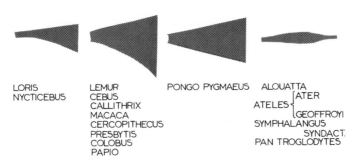

LORIS   LEMUR   PONGO PYGMAEUS   ALOUATTA
NYCTICEBUS   CEBUS   ⌈ATER
                     CALLITHRIX   ATELES⌐
                     MACACA   ⌊GEOFFROYI
                     CERCOPITHECUS   SYMPHALANGUS
                     PRESBYTIS   SYNDACT.
                     COLOBUS   PAN TROGLODYTES
                     PAPIO

Fig. 2. Main types of shape of the long head of the biceps femoris, i.e., ischiocrural lateral muscle

studied material, similar relations were noted only in *Alouatta*. However, the main insertion of the biceps muscle in *Symphalangus* remains on the tibia, the fibular insertion being secondary. Figure 2 shows the main types of shape of the long head of the biceps femoris muscle in the studied Prosimiae and monkeys.

The Tenuissimus Muscle

Among the lower limbs of 60 primates of various genera and species examined, including 15 platyrrhine monkeys, the tenuissimus muscle was present only in the latter.

In the majority of the investigated genera, the tenuissimus muscle coexisted with the long head of the biceps femoris muscle (i.e., the lateral ischiocrural muscle), namely in the genera: *Cebus, Callithrix, Saimiri* and *Leontocebus,* in which the short head of the biceps femoris was absent.

In two genera of the platyrrhine monkeys, *Ateles* and *Alouatta,* both heads of the biceps muscle were present and the tenuissimus muscle was absent. This situation was met in all eight examined limbs of *Ateles* and *Alouatta.*

The tenuissimus muscle in different genera was similar, its dimensions depending on the size of the monkey. This muscle is a narrow strand running on the medial surface of the lateral ischiocrural muscle (Fig. 3), its width ranging from 1/2 mm (in *Callithrix* and *Saimiri*) to 2–3 mm (in *Cebus*). It begins as a slender tendon arising from the fascia on the medial surface of the superficial gluteus muscle (caudal part).

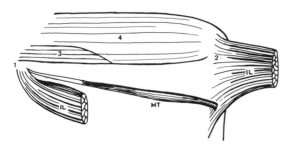

Fig. 3. Tenuissimus muscle in *Cebus capucinus. IL,* ischiocrural lateral muscle (transected and deflected); *MT,* tenuissimus muscle; *1,* ischial tuberosity; *2,* lateral condyle of the tibia; *3,* superficial gluteus muscle; *4,* quadriceps femoris muscle

The tendon often divides into 2, 3, or 4 branches at first. In four *Cebus* and one *Callithrix* this tendon also received strands from the first and the second coccygeal vertebrae. In one *Cebus*, initial tendinous strands arose also from the ischial tuberosity. Running on the medial surface of the lateral ischiocrural muscle, the tenuissimus muscle crosses it obliquely in the downward and foreward direction, emerging in its distal part and forming its lower border. The terminal insertion is the passage of muscle fibers into the crural fascia. Here, the tenuissimus muscle forms the lower part of the terminal insertion of the lateral ischiocrural muscle.

In our material, the tenuissimus muscle was inconstant in *Cebus, Callithrix,* and in *Leontocebus.* In one *Cebus* it was absent on both sides, and in another specimen on one side. Bilateral absence of the muscle was observed also in one *Callithrix.*

In all, the tenuissimus muscle was absent in 7 of the 22 examined extremities (Prejzner-Morawska 1977).

According to the literature, the tenuissimus muscle is not constant. According to Hill (1960), it is poorly visible in *Callithrix* and *Leontocebus,* and is often mistaken for a nerve. Klaatsch (1911) reported that the tenuissimus muscle is 1/2 mm wide, and is often poorly visible. This may be the reason why Beattie (cited after Uhlman 1968) failed to find the tenuissimus muscle in 26 extremities of Callithricidae.

According to most authors (Hill 1953–1960, Uhlman 1968) the initial insertion is on the fascia under the superficial gluteal muscle, and, although not invariably, on the first and the second coccygeal vertebrae. None of the authors mentions tendinous strands arising from the ischial tuberosity, such as were observed in this study in one *Cebus.*

Our results confirm many observations of many other investigators, indicating that in primates the tenuissimus muscle occurs only in some of the platyrrhine monkeys *(Cebus, Callithrix, Saimiri, Leontocebus),* although not as a rule. Coexistence of the tenuissimus muscle with the short head of the biceps femoris muscle does not occur, indicating common origin of both muscles.

The Semitendinosus Muscle

In 9 Prosimiae and 15 platyrrhine monkeys the semitendinosus muscle extends from the ischial tuberosity to the tibia. The ratio of the tendinous to the muscular part, however, is different from that in Man. The tendinous part constitutes one-third (in lemurs) to one-fourth (in platyrrhine monkeys) of the whole length of the muscle. In some platyrrhines *(Callithrix penicillata, Saimiri sciureus, Leontocebus rosalia)* the semitendinosus muscle has an accessory, less strongly developed, initial insertion on the first caudal vertebra.

The terminal insertion in all animals was on the proximal part of the anterior crest of the tibia, below its tuberosity, through a common tendon with the gracilis muscle. The degree of fusion of the tendons of the semitendinosus and gracilis muscles varies in different genera. from only adjacence to very strong adhesion.

In *Nycticebus coucang* and in one *Lemur varius,* an interflexor muscular bundle running from the semitendinosus to the lateral ischiocrural muscle was remarked upon (Prejzner-Morawska 1978).

Triceps Surae Muscle

Triceps surae muscle, in primates, similarly as in Man, consists of the gastrocnemius and soleus muscles; its tendon is the tendo calcaneus, inserting into the posterior or postero-inferior surface of the calcaneus.

Gastrocnemius Muscle

This arises by two proximally independent heads. Their initial insertion is situated above the corresponding condyles of the femur. In the initial tendon of both heads the seamoid bone is found. Only in Anthropomorpha was it not observed.

The level of fusion of both heads of gastrocnemius varies in different genera and species. Usually in most Prosimiae and monkeys the union takes place a little above the middle of the leg (Fig. 4).

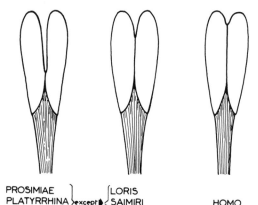

PROSIMIAE ⎫   ⎧LORIS   **Fig. 4.** Level of union of both heads of
PLATYRRHINA ⎬except ⎨SAIMIRI   HOMO   the gastrocnemius muscle in studied
CATARRHINA ⎭   ⎩SYMPHALANGUS   monkeys

In *Loris* among Prosimiae, *Saimiri* among Platyrrhina, and in *Symphalangus* the fusion is placed higher, approximately in the lower four-fifth of the leg. According to Frey (1913) the highest union of both heads can usually be found in Man.

Figure 5 illustrates the proportions of both heads of gastrocnemius and their downward extent. In *Presbytis entellus* the muscular belly of the lateral head is larger and its downward extent is greater. The muscular belly of the medial head was in all eight chimpanzees examined more strongly developed than that of the lateral head, the difference being plainly visible and pertaining to the breadth as well as thickness of the muscle and its distal extent (so-called downward spread according to Różycki 1922). As a result of the low descent of the muscle fibers, the terminal tendon of the gastrocnemius muscle is very short in chimpanzees (Urbanowicz and Prejzner-Morawska 1972). Extremely different gastrocnemius is shown in Man to emphasize also the differences of length of the tendinous part. As we know, in Man the tendon is long, reaching 65% of the whole muscle length (Frey 1913).

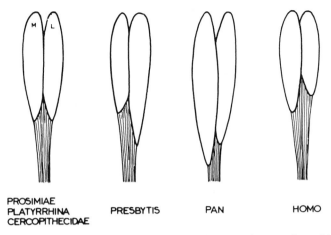

PROSIMIAE
PLATYRRHINA   PRESBYTIS   PAN   HOMO
CERCOPITHECIDAE

Fig. 5. The gastrocnemius muscle. Relation between the muscular and tendinous part and the downward extent of the muscular bellies. *M*, medial head; *L*, lateral head

The Soleus Muscle

The soleus muscle in lemurs, Platyrrhina, and Catarrhina arises solely from the back of the head of the fibula (Fig. 6). In *Loris , Nyticebus, Ateles,* and chimpanzee an accessory (in some cases relatively large) insertion to the shaft of the fibula was observed. The tibial insertion in a primitive form occurs in two of eight chimpanzees examined. According to Rózycki (1922) it is much more frequent. This insertion is well developed and constant only in Man.

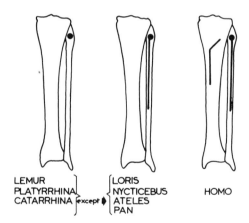

LEMUR   ⎤   ⎡LORIS
PLATYRRHINA   ⎥   ⎨NYCTICEBUS   HOMO
CATARRHINA {except ♦} ⎥   ⎥ATELES
               ⎦   ⎩PAN   **Fig. 6.** Initial insertion of soleus muscle

The soleus muscle in Prosimiae and lower monkeys is narrow, thin, and fusiform (Fig. 7). In Anthropomorpha it is well developed, resembling in shape the analogous muscle in Man.

Figure 8 shows the level and degree of union between the soleus and gastrocnemius muscle. In Catarrhina these muscles are most often independent and do not fuse with

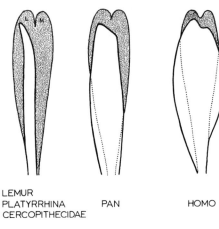

Fig. 7. Shape of soleus muscle

LEMUR
PLATYRRHINA   PAN   HOMO
CERCOPITHECIDAE

CATARRHINA

NYCTICEBUS
LEMUR
CEBUS
PAN

HOMO
LORIS
ATELES
HYLOBATES

Fig. 8. Degree and level of union between gastrocnemius and soleus muscles. *G,* gastrocnemius muscle; *S,* soleus muscle

each other. Sometimes the area of their adhesion is very small, just above the calcaneus. In Prosimiae, Platyrrhina, chimpanzee, and *Pongo* the union is strong and relatively large. The upper level of the fusion takes place a little beneath the middle of the leg. The high level of the union is characteristic of Man. In *Loris, Ateles,* and *Hylobates* the level of union is similar to Man.

The adhesion of the gastrocnemius with the soleus begins from their lateral margins. The upper line of adhesion runs down obliquely to the medial margin of both muscles. Only in *Ateles* does the soleus fuse with the medial head of gastrocnemius.

Figure 9 illustrates the shape and size of the area of union of the gastrocnemius with the soleus muscle.

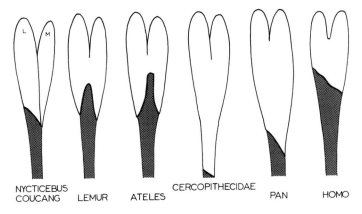

Fig. 9. Area of union of the gastrocnemius and soleus muscles

The Tendo Calcaneus

This is the common tendon of the gastrocnemius and soleus:

1. In Cercopithecidae it does not occur because both muscles independently reach the distal insertion.

2. In *Ateles* and in chimpanzee it is of predominantly "muscular" character: the tendon of gastrocnemius receives many fleshy fibers from the soleus, almost to its lower end.

3. In the remaining Prosimiae and monkeys this tendon resembles the human one, but it is not so strong and thick.

4. Tendo calcaneus (Achillis) in classic form occurs only in Man.

References

Frey H (1913) Der Musculus triceps surae in der Primatenreihe. Morphol Jahrb 47:1–191

Hill WCO (1953–1960) Primates. Comparative anatomy and taxonomy, vol I: Strepsirhini, vol II: Haplorhini: Tarsioidea, vol III: Pithecoidea: Platyrrhini, vol IV: Cebidae, part A. Univ Press, Edinburgh

Klaatsch H (1911) Über eine dem Tenuissimus ähnliche Variation am Biceps femoris des Menschen. Anat Anz 38:305–310

Loth E (1931) Anthropologie des parties molles. Mianowski-Masson et Cie, Varsovie Paris

Prejzner-Morawska A (1977) The tenuissimus muscle in primates. Folia Morphol (Warsaw) 36: 135–140

Prejzner-Morawska A (1978) The semitendinosus muscle in lemurs and platyrrhine monkeys. Folia Morphol (Warsaw) 37:103–112

Prejzner-Morawska A, Urbanowicz M (1971) The biceps femoris muscle in lemurs and monkeys. Folia Morphol (Warsaw) 30:465–482

Różycki S (1922) Morphology of the muscular system in chimpanzees. Gebethner-Wolf, Poznan (in Polish)

Uhlman K (1968) Hüft- und Oberschenkelmuskulatur. Systematische und vergleichende Anatomie. In: Hofer H, Schultz AH, Starck D (eds) Primatologia, vol IV, Lief 10, Karger, Basel New York

Urbanowicz M, Prejzner-Morawska A (1972) The triceps surae muscle in chimpanzees. Folia Morphol (Warsaw) 31:432–440

Morpho-Functional Analysis of the Articular Surfaces of the Knee-Joint in Primates

C. Tardieu [1]

Variability of the Mammal Knee-Joint

Figure 1 exhibits the epiphysis of the distal femur, representing each of the mammal locomotor patterns:
— the unguligrade, illustrated by the horse,
— the digitigrade, illustrated by the panther,
— the plantigrade, illustrated by the bear.
We can observe two significant morpho-functional features:

The Morphology of the Femoral Trochlea. Passing from the unguligrade, through the digitigrade to the plantigrade, one sees a gradient of increasing flatness of the femoral trochlea; the trochlear groove becomes shallower and broader.

The functional meaning of this feature is the following: in unguligrades the patella fits into this deep groove of the trochlea, permitting a very well-fitted, highly channeled gliding movement that facilitates very rapid motion. In the run of the horse, movements are exclusively parasagittal, involving only flexion-extension of the knee and excluding rotation. In plantigrade, conversely, this tight-fitting structure does not occur; instead the patella moves freely on the trochlear surface, resulting in a less restricted functioning of this joint.

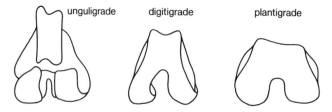

unguligrade   digitigrade   plantigrade

Fig. 1. Inferior epiphysis of a left femur

1   Laboratoire d'Anthropologie du Musem National d'Histoire Naturelle, Musée de l'Homme, place du Trocadéro, 75116 Paris, France

The Morphology of the Zone of Articular Transition Between the Trochlea and Condyles. In unguligrades there is a break of articular continuity, while in digitigrades and plantigrades one sees an uninterrupted articular surface. This area corresponds precisely to the zone of femoro-tibial contact in full extension of the knee. This full extension is characteristic of plantigrades. The unguligrade break makes complete extension of the joint impossible. Even in an upright position the knee of an unguligrade mammal is obviously flexed. In all living primates there is an articular continuity between trochlear and condylar surfaces.

Comparative Development of the Femoral Trochlea and Condyles in Primates

I have more precisely studied these articular surfaces of the knee in primates. My study was performed on a large sample of primates, representing the principal systematic groups and the most diverse and characterstic locomotor and postural behaviors.

In the literature one found only linear measurements of these surfaces, which are not very significant from a functional point of view. I chose to measure the curvilinear length, the developed length of trochlea and each condyle (Fig. 2). It permits a real appreciation of the actual displacement of the femur on the tibia and of the patella on the trochlea in flexion-extension and rotation movements. These measurements prove to be very significant in the discrimination of the primate knee joint.

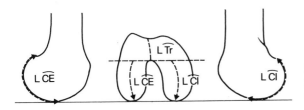

Fig. 2. Measurements of curvilinear lengths of the femoral trochlea and condyles

In Fig. 3, I show the relationship between the curvilinear development of the two condyles and trochlea. The X axis represents the index of developed length of the internal condyle over the developed length of the trochlea. The Y-axis represents the index of developed length of the external condyle over the developed length of the trochlea. Along the slanting line 00' are plotted the points which represent the equality of the two condyles. In the region to the right are plotted the points which represent a greater development of the internal condyle over that of the external. Moving up the bissector, the developed length of the condyles increases. Moving down the bissector, the developed length of the trochlea increases. Point Ω represents the equality of the three factors. Generally, the lighter animals are plotted toward the bottom of the graph, with preponderance of the trochlea, and the heavier animals are plotted toward the top of the graph, with preponderance of the condyles.

All the "prosimians" are situated below the index 130, excepting the Lorisines, slow-climbers.

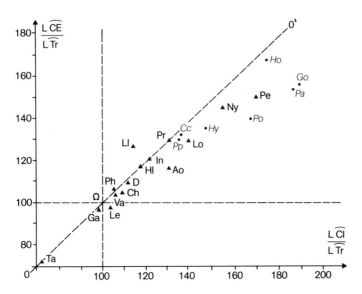

Fig. 3. Distribution of the different genera of primates, according to the curvilinear length *(L)* of the trochlea *(Tr)*, of the internal condyle *(CI)* and of the external condyle *(CE)*. *X axis: L CE/L Tr* ratio. *Y axis: L CI/L Tr* ratio. Ω: Point corresponding to the equality of the trochlea, of the internal condyle and of the external condyle. An average for each genus of primates.
Ao, Aotus; Cc, Cercopithecus; Ch, Cheirogaleus; D, Daubentonia; Ga, Galago; Go, Gorilla; Hl, Hapalemur; Ho, Homo; Hy, Hylobates; In, Indri; Le, Lemur; Ll, Lepilemur; Lo, Loris, Ny, Nycticebus; Pa, Pan; Pe, Perodicticus; Ph, Phaner; Po, Pongo; Pp Papio; Pr, Propithecus; Ta, Tarsius; Va, Varecia

Relation Between Trochlea and Condyles. In the best leapers, the development of the trochlea is maximal in relation to the condyles, and is frequently accompanied by the presence of a superior patella. This superior patella fits tightly in the trochlear groove in the movement of hyperflexion of the knee, which immediately precedes the leap: for example in *Tarsius.*

Relation Between the Two Condyles. In primates, for the most part, the developed lengths of the two condyles are roughly equal. However, we do see some variation, again for Lorisines, and strongly in the pongids which show a greater developed length of the internal condyle. This corresponds to the performance of combined rotation: this is when rotation and flexion-extension movements occur at the same time in the joint, introducing a screw-like mechanism.

Morpho-Functional Variation of the Catarrhine Knee-Joint

I restricted my sample and focused only on the catarrhine primates. This resulted in a large morpho-functional table comparing the knee joint of cercopithecids, pongids, and modern Man. Only some parts of it are presented here.

Traditional Features

The first section of Fig. 4 shows three functionally linked features: (1) Degree of femoral obliquity, (2) Form of the femoral trochlea, and (3) Lateral profile of the external condyle. They are "traditional" features.

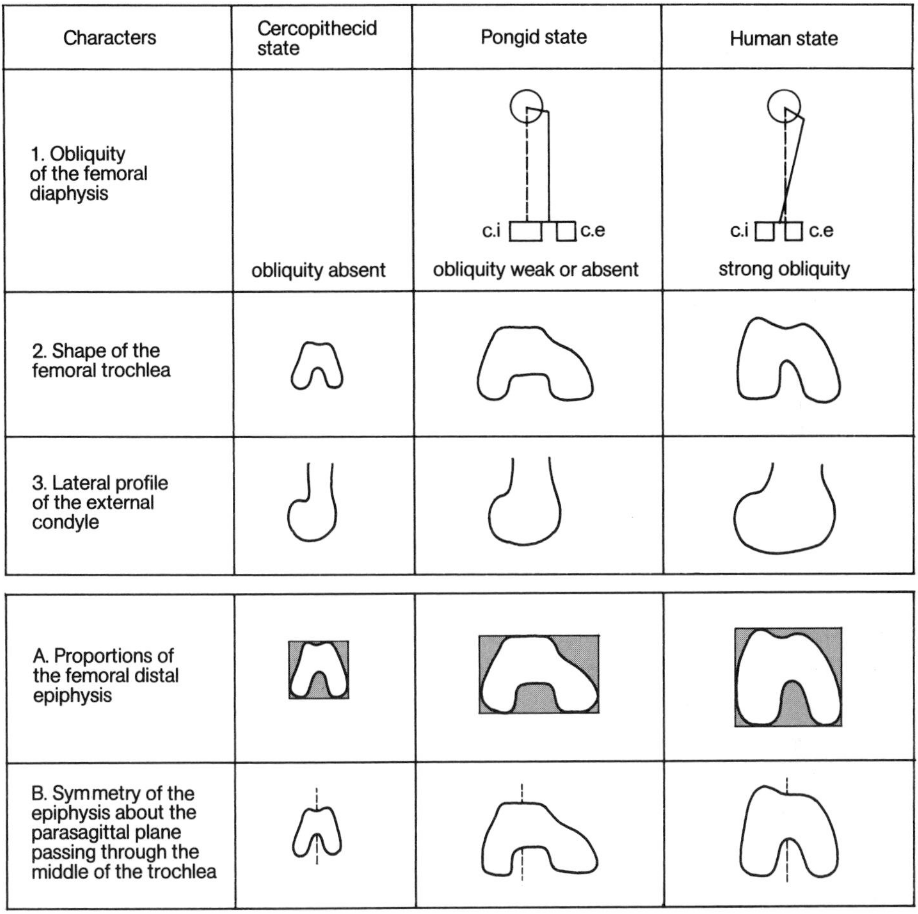

Characters	Cercopithecid state	Pongid state	Human state
1. Obliquity of the femoral diaphysis	obliquity absent	c.i ☐☐ c.e obliquity weak or absent	c.i ☐☐ c.e strong obliquity
2. Shape of the femoral trochlea			
3. Lateral profile of the external condyle			
A. Proportions of the femoral distal epiphysis			
B. Symmetry of the epiphysis about the parasagittal plane passing through the middle of the trochlea			

Fig. 4. Variation of the cercopithecid, pongid, and human knee joint

Degree of Femoral Obliquity

The position of the femoral diaphysis functionally determines the direction of the line of the load of the lower member in relation to the two condyles.

The absence of femoral obliquity in cercopithecids and pongids causes the line of load of the lower member, originating from the femoral head, to pass either between the two condyles (cercopithecids) or through the internal condyle (pongids). The strong obliquity of the femur in Man is one of the elements which permits the line of load to be disposed toward the external condyle. This causes the extended articulation of the knee in Man to be placed in a position more internal to the hip.

Form of the Femoral Trochlea

Only slightly hollowed in the cercopithecids and flat in the pongids, the trochlea in Man exhibits a deep groove and an external lip that is higher and anteriorly more projecting. This character in Man is due to the obliquity of the femoral diaphysis. It permits, in effect, the maintenance of the medio-lateral stability of the patella in full extension of the knee (Heiple and Lovejoy 1971).

Lateral Profile of the External Condyle

Circular in the cercopithecids and the pongids, it is elliptic in Man. In functional terms, this elliptical contour corresponds to an increase of the radius of curvature of this condyle in its inferior part. Consequently, the radius of curvature is increased and maximized precisely in the femoro-tibial contact zone corresponding to the extension of the knee. By contrast, the other primates display a circular projection of the condyle (except the internal condyle of pongids), that does not favor a position of full extension of the knee joint (Heiple and Lovejoy 1971).

Thus, the abducted and flexed position of the knee is characteristic of pongids, while the adducted and extended position of the knee is characteristic of Man. The adduction of the human knee places this intermediate articulation in a medial position in relation to the hip joint. This permits the knee and ankle joints to be placed almost directly under the center of gravity during the single support phase of gait, one of the characteristics essential to the efficiency of human bipedal locomotion.

In Man, the support polygon being smaller, the static support of the body is obviously less efficient; however, during the successive disequilibria of the body during the single support phases of locomotion, the dynamic support of the body is more efficient: this is because the load transmitted by the lower member approaches more closely the line of gravity of the body.

Other Discriminant Features

(1) Proportions of the distal femoral epiphysis, (2) Symmetry of the distal femoral epiphysis about the parasagittal plane passing through the middle of the trochlea, and (3) Difference between the width of the femoral intercondylar notch and the interspinal distance of the tibia.

The first two features are displayed in the second section of Fig. 4. The third feature is depicted on Fig. 6. These features have not been described earlier and could be opposed to the "traditional" features.

Proportions of the Distal Femoral Epiphysis

The inferior view of the epiphysis is very significant from a morpho-functional point of view. To express the proportions of the epiphysis, I calculated the index: $\frac{\text{length of the external condyle}}{\text{posterior width of the epiphysis}}$, that we have used on the ordinate axis of Fig. 5.

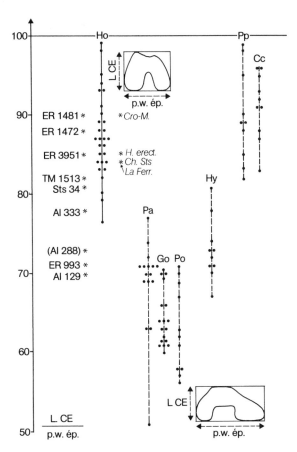

Fig. 5. Proportions of the distal femoral epiphysis. *L CE,* Linear length of the external condyle; *p. w. ep.,* posterior width of the epiphysis. The symbols * represent the fossil hominids. *Cc, cercopithecus; Go, Gorilla; Ho, Homo; Hy, Hylobates; Pa, Pan; Po, Pongo; Pp, Papio*

Length of the external condyle was chosen as the linear length of the condyle, from a lateral view perpendicular to the axis of the diaphysis and to the middle height of the condyle. Posterior width of the epiphysis was taken from a posterior view at the middle height of the condyles.

It is concluded from Fig. 5 that the posterior width of the epiphysis is always greater than the length of the external condyle, for no primate is above index 100, and that three different states are seen for this character:

The human state, which resembles the cercopithecid: the epiphysis presents a posterior width slightly greater than the length of the external condyle. The epiphysis is inscribed within a square. The fact that Man and the cercopithecids present a square epiphysis in a frontal view does not signify that the overall forms are the same. The distal epiphysis in these two groups are distinguished by the profile of the external condyle (circular in cercopithecids, elliptical in Man), the morphology of the internal condyle, and the absolute dimensions of the epiphysis contained within the square, which are much smaller in cercopithecids.

The pongid state: the posterior width of the epiphysis is one and a half to two times greater than the length of the external condyle; we shall say, schematically, that the epiphysis is inscribed within a rectangle.

The hylobatid state which is intermediate between the two preceding states. There is no overlap between the state realized in Man and the state realized in pongids. In Man, we observe an antero-posterior lengthening of the articular surfaces, while in pongids we observe a mediolateral increase in width of the distal femoral epiphysis. We think that these two states are diametrically opposed.

Symmetry of the Distal Epiphysis About the Parasagittal Plane Passing Through the Middle of the Trochlea (Character B on Fig. 4)

The "square" epiphysis of cercopithecids is very symmetrical; there is symmetry of the trochlea of the two condyles, of the intercondylar notch. I would consider it an undifferentiated morphology. Conversely, the "rectangular" pongid structure is strongly asymmetrical, with preponderance of the internal side over the external.

The human "square" structure shows a slight, but very different, asymmetry due to the curvature of the anterior segment of the internal condyle, used in the terminal rotation mechanism.

Note the shape of the intercondylar notch in each case; symmetrical and arch shaped in cercopithecids, wider than it is high in pongids and following the rectangular shape, higher than it is wide in Man, and curved on tis internal wall, following the anterior curvature of the internal condyle.

Difference Between the Width of the Femoral Intercondylar Notch and the Interspinal Distance of the Tibia

We measured: (1) the width of the intercondylar notch, taken at mid height of the notch, the femur lying on an horizontal plane, and (2) the interspinal distance of the tibia: the distance between the most proximal points of the two tibial spines.

We calculated the index $\dfrac{\text{interspinal distance of the tibia}}{\text{width of the intercondylar notch}}$, that we plotted on the ordinate axis of Fig. 6.

Figure 6 shows the influence of the weight and of the locomotor pattern on this character. For the most part, the lightest primates present the greatest difference in width, while the heaviest primates exhibit a smaller difference between interspinal distance and notch width.

This graph shows that the cercopithecids exhibit a strong difference between the width of the tibial spines and of the intercondylar notch. Conversely, Man shows the least difference between these two widths. The different pongid genera do not strictly overlap and offer strong variation. This variation is perhaps explained by the influence of the body weight. In fact, in the chimpanzee, the lightest pongid, one sees the greatest difference in width; in the heavier orang-utan the difference is less; and finally, in the gorilla the difference in width is the smallest. The gibbon shows the strongest difference between these widths.

This feature concerns the rotation movements of the knee joint. Thus, to facilitate the interpretation we must first distinguish the movement of "combined rotation"

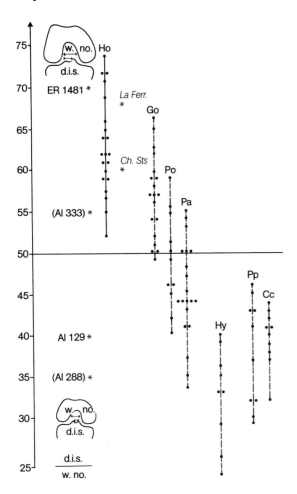

Fig. 6. Difference between the width of the intercondylar notch of the femur and the interspinal distance of the tibia. *d. i. s.*, distance inter spines; *w. no,* width of the intercondylar notch

and the movement of "independent rotation", each of which takes part in the functioning of the knee joint:

Combined Rotation. As noted in the first part of this paper, the movement of combined rotation is the consequence of the unequal curvilinear length of the two femoral condyles, that of the internal condyle always being greater (Fig. 3). Very developed in pongids, it presents a weaker amplitude in Man. While rotation is "combined" with flexion in pongids, in Man it is invariably combined with extension. In fact, the terminal automatic rotation movement, characteristic of the human knee joint, which locks the knee at the end of the extension, is a part of this combined rotation movement.

Independent Rotation. While combined rotation is an obligatory movement, a necessary consequence of the geometry of the articular surfaces, independent rotation represents pure rotation movement, which can only be performed by a certain degree of flexion. It introduces a certain laxity into the articulation of the knee.

In the feature described in Fig. 6, the tibial spines represent the pivot of rotation of the femur on the tibia; the difference of width between this tibial pivot of rotation and the femoral intercondylar notch gives two indications: (1) it reflects the amplitude of the movement of independent rotation of the knee; (2) it also indicates whether the combined rotation movement is guided in a very relaxed or in a very precise fashion.

Consequently, the primates of our sample offer very significant variability for this character: this figure shows an independent rotation maximum in the gibbon, strong in cercopithecids, variable in the pongids as function of weight (rotation decreasing as weight increases), and weak in Man. This graph indicates that combined rotation is performed with a very lax guidance in the gibbon, a variable guidance in the pongids (guidance increasing with increasing body weight), and with precise guidance in Man.

It appears that, while weight is a significant factor, the mode of locomotion very clearly influences this character: Man, although much lighter than gorilla, exhibits a closer fit of the tibial spines in the intercondylar notch. This very precise fit of the articular elements in Man would be a feature linked to his exclusive erect and bipedal locomotion. In pongids, closest fit of the joint is found in the gorilla, the heaviest, but also the most terrestrial of the pongids.

Independent rotation displays a very weak amplitude in Man. Bugnion (1892) showed that it is an essentially passive movement in Man. We suggest that the relatively greater amplitude of independent rotation observed in nonhuman primates of our sample, represents an active movement rather than just a passive one.

The very close fit of the tibio-femoral articular surfaces suggests that joint solidity has been emphasized at the expense of joint laxity in Man. This disposition can be clearly linked to the foot's role in support of the body, to the exclusion of any great prehensile ability, inherent in bipedally erect human locomotion.

The loose fit of the articular surfaces in cercopithecids, hylobatids, and light pongids and the consequent laxity of the knee joint signify that the leg and the foot can be placed on the substrate in a much freer fashion than in Man and that the mobility and prehensility of the foot are greatly complemented.

Implications for the Interpretation of Fossil Hominids

The interpretative table of catarrhine primates is now applied to the fossil hominids. The fossils are presented on Figs. 5 and 6. Very clearly, Pleistocene fossils had definitively achieved the modern structure of the human knee. The significant area for locomotor evolution is situated before *Homo erectus,* precisely among the Plio-Pleistocene fossils and earlier.

Features Common to Fossil Hominids

All the fossils present the three following characters: (1) femoral obliquity, (2) deep patellar groove, with higher lateral lip, and (3) elliptical profile of the external condyle. These three characters distinguish the hominids from the nonhominid fossils.

Features Discriminating Between Fossil Hominids

However, fossil hominids are divided into two groups when the interpretation of the "square" or "rectangular" epiphysis is applied. The first group of fossils overlaps human variation, and the second group overlaps the pongids in the upper limit.

These two groups of fossils are shown in Fig. 7. Only one big-sized fossil and one small-sized fossil have been drawn as representative of each group. In parenthesis are listed the fossils included in the same group as the representative fossils.

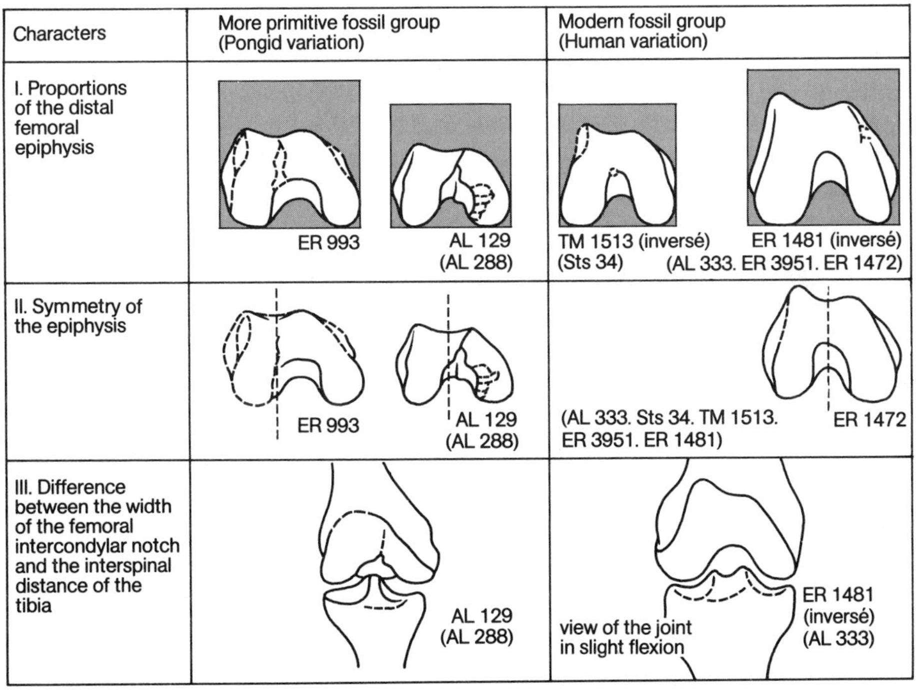

Fig. 7. Distinctive features of the two groups of fossil hominids

On the right, the fossils of the modern group are square-shaped. The epiphysis is almost symmetrical about the parasagittal plane passing through the middle of the trochlea. This "square" group displays a very close-fitting joint. Here the solidity of the joint would restrict the mobility of the foot and of the ankle joint, as in modern Man.

On the left, the fossils of the other group are rectangular-shaped. The epiphysis is asymmetrical about the parasagittal plane passing through the middle of the trochlea, with slight preponderance of the internal side. Here, the loose fit of the tibial spines in the femoral intercondylar notch allows for greater rotation in the joint and greater mobility of the foot and of the ankle, such as we suggest. In our point of view, the fossils of the second group would represent a significantly more primitive condition.

On the right, the fossil KNMER 1481, from East Turkana, includes a complete femur and a complete tibia. The knee joint is very well preserved and is the best representation of the modern group, *Homo cf. habilis.* This group is well represented by other East Turkana material: KNMER 1472, complete femur, and KNMER 1951, distal femur.

The next fossil, TM 1513, from Sterkfontein, is an isolated distal femur, belonging to the modern group.

The fossil AL 129, from Afar, is a complete very well preserved knee joint, from the second group. The knee joint AL 288, from Afar, "Lucy's" knee joint, is exactly of the same type as AL 129.

Notice that the two femurs, AL 129 and TM 1513, are of the same small size, but of different morphologies, in my point of view.

These Afar fossils have been attributed to the new species *Australopithecus afarensis* (Johanson et al. 1978). Some authors include in the variation of this new species the fossil AL 333-4. Our study would incline to place this critical fossil in the first modern group.

On the left, the fossil KNMER 993, from East Turkana, is an isolated distal femur.

Conclusion

The study of the articular surfaces of the knee joint shows that these surfaces are excellent indicators of the different modes of functioning of this articulation involved in the diverse locomotor and postural behaviors of primates.

Each part of this paper corresponds to a more detailed study (Tardieu 1979a,b,c), of which only a few elements have been presented. The measurements of *developed lengths* of the femoral trochlea and condyles prove to be very significant in the discrimination of primate knee joint:

1. The greater curvilinear length of the trochlea in relation to that of the condyles corresponds to great amplitude of flexion and extension and is often accompanied by the presence of a *superior patella.* This morphology characterizes the prosimians, excepting Lorisines (slow climbers), and opposes prosimians to the other primates. The development of the trochlea is maximum in the lightest prosimian leapers.

2. The greater curvilinear length of the internal condyle over that of the external condyle causes rotation movement to be combined with flexion-extension movement. This permits a *screw-like mechanism* in the femoro-tibial articulation, varying in development in the primates, with maximum development in pongids.

In catarrhine primates, we distinguish three morpho-functional types of joint: cercopithecid, pongid, and human. We suggest that the very symmetrical distal femoral cercopothecid epiphysis corresponds to an undifferentiated type, while pongid and human knee joints are specialized in two very different ways.

When applied to fossil hominids, our analysis of catarrhine knee joints shows that all the fossil hominids display the following "traditional" human features: (1) femoral obliquity, (2) deep trochlear groove with higher lateral lip, and (3) elliptical profile of the external condyle. However, fossil hominids can be divided into two groups, when other features are applied:

The modern group which exhibits:

1. square-shaped epiphysis, corresponding to an anteroposterior lengthening of the distal femoral articular surfaces, characteristic of modern Man,

2. close fit of the tibial spines in the femoral intercondylar notch, characteristic of modern Man; it implies a very weak amplitude of independent rotation: here, joint solidity has been emphasized at the expense of joint laxity, as in modern Man.

This group includes some Plio-Pleistocene fossils and all the later hominids: the complete knee joint KNMER 1481, from East-Turkana, is the most representative fossil of this group.

The more primitive group which exhibits:

1. A rectangular-shaped and asymmetrical distal epiphysis, corresponding to a medio-lateral broadening of this femoral epiphysis that one can observe in pongids.

2. A looser fit of the tibial spines in the intercondylar notch, allowing for greater rotation in the joint, as in some pongids; here, the laxity of the knee joint is significantly more developed than in the modern group.

The complete knee joint AL 129, from Afar, is the most representative fossil of this group.

Acknowledgments. I would like to thank R.E.F. Leakey, Nairobi, and Dr. D.C. Johanson, Cleveland, for their kind permission to study the original fossils in their care; Prof. Anthony, Paris, Dr. Y. Coppens, Paris, for access to their primate and human collections and for practice of primate dissections; Prof. A Delmas and Dr. P Le Floch for practice of human dissections.

I am also grateful to the International Afar Research Expedition, the Koobi Fora Research Expedition, and field expedition to Sterkfontein. This work was supported by R.C.P. 292 (C.N.R.S.).

References

Bugnion E (1892) Le mécanisme du genou. Thèse de médecine. In: Rec Inaugural Univ Lausanne, pp 333–374

Day HM (1978) Functional interpretations of the morphology of postcranial remains of Early African hominids. In: Jolly CI (ed) Early hominids of Africa. Duckworth London, pp 311–345

Grassé PP (1967) Traité de zoologie, anatomie, systématique, biologie. T XVI Fasc 1. Mammifère, téguments, squelette. Masson, Paris

Halaczek B (1972) Die Langknochen der Hinterextremität bei Simischen Primaten. Juris Druck und Verlag, Zürich, 177 pp

Heiple KG, Lovejoy CO (1971) The distal femoral anatomy of Australopithecus. Am J Phys Anthropol 35:75–84

Johanson DC, Coppens Y (1976) A preliminary anatomical diagnosis of the first Plio-Pleistocene discoveries in the Central Afar, Ethiopia. Am J Phys Anthropol 45:217–233

Johanson DC, White TD (1979) A systematic assessment of Early African hominids. Science 203: 321–330

Johanson DC, White TD, Coppens Y (1978) A new species of the genus *Australopithecus* (Primates: Hominidae) from the Pliocene of Africa. Kirtlandia 28:1–14

Kapandji IA (1970) Physiologie articulaire. Membre inférieur. Fasc II. Maloine, Paris, 234 pp

Leakey REF (1971) Further evidence of Lower Pleistocene Hominids from East-Turkana, Kenya. Nature (London) 231:241–245

Leakey REF (1972) Further evidence of Lower Pleistocene Hominids from East-Turkana, North Kenya. Nature (London) 237:264–269

Leakey REF (1973) Further evidence of Lower Pleistocene Hominids from East-Turkana, North Kenya. Nature (London) 242:170–173

Plas F, Viel E (1975) La marche humaine. Kinesiologie dynamique, biomécanique et patho-mécanique. Masson, Paris

Poirier P (1866) Contribution à l'anatomie du genou. Prog Méd 357–371

Rainaut JJ, Lotteau J (1974) Télémétrie de la marche, goniométrie du genou. Rev Chir Orthoped 60:97–107

Retterer E, Vallois H (1912) De la double rotule de quelques primates et de quelques rongeurs. C R Soc Biol, Oct

Rouvière H (1970) Anatomie humaine descriptive et topographique, Tome II: Membres, système nerveux central. Masson, Paris

Sonnenschein A (1951) Die Evolution des Kniegelenkes innerhalb der Wirbeltiere. Acta Anat 13: 288–328

Steindler A (1955) Kinesiology of the human body under normal and pathological conditions. CC Thomas Publ, Springfield, 709 pp

Tardieu Ch (1979a) Analyse morpho-fonctionnelle de l'articulation du genou chez les Primates. Application aux Hominidés fossiles. Thèse Sci Nat (Paléontologie humaine), Univ Paris VI

Tardieu Ch (1979b) Aspects biomécaniques de l'articulation du genou chez les Primates. Bull Soc Anat Paris 4:66-86

Tardieu Ch, Jouffroy FK (1979c) Les surfaces articulaires du genou chez les Primates. Etude préliminaire. Ann Sci Nat Zool 1/23–28

Vallois H (1914) Etude anatomique de l'articulation du genou chez les Primates. Impr Coop Ouvrière «L'Abeille»' Montpellier, 467 pp

Vallois H (1919) L'épiphyse inférieure du fémur chez les Primates. Bull Mem Soc Anthropol Paris 10:21–45, 80–107

Outlines of the Distal Humerus in Hominoid Primates: Application to Some Plio-Pleistocene Hominids

B. Senut [1]

Considering hominid evolution (and especially Plio-Pleistocene hominids), it is striking that the forelimb, and especially the humerus, has been studied only by isolated authors (Broom et al. 1950, Day 1976, 1978, Day et al. 1976, McHenry 1973, McHenry and Corruccini 1975, Oxnard 1973, 1975, Robinson 1972, Straus 1948). The skulls and teeth were probably considered as more interesting, and usually used in reconstructions of human phylogeny. However, recent discoveries realized in East Africa (East Turkana in Kenya, Hadar in Ethiopia) have provided a large sample of distal humeri which permit the differentiation of three main groups among Plio-Pleistocene hominids (Senut 1978).

These new specimens show that the humerus of early Man, often considered as equivalent to that of modern humans, is, in fact, different, and some of them share features with *Homo sapiens* or pongids. The first group includes the specimens of Kanapoi (KNM KP 271) in Kenya and from Melka Kunture (Gombore IB 7594) in Ethiopia. Their features are similar to modern Man: poorly salient *epicondylus lateralis*, poorly distally developed *capitulum humeri*, weakly developed lateral crest of the *trochlea humeri*. I called these specimens *Homo sp.* The second group is divided into two subgroups. The first one, partly characterized by the marked anteroposterior flattening of the distal shaft, a strong lateral salience of the *epicondylus lateralis*, the proximo distal elongation of the *capitulum humeri*, and a shallow *fossa olecrani* set in the middle of the biepicondylar width, includes KNM ER 739, KNM ER 740, KNM ER 1504, KNM ER 3735 from East Turkana in Kenya, and TM 1517 from Kromdraai in South Africa. The second subgroup consists of most of the Afar material and is remarkable by the gracility of the bones, the "double-trochlea" pattern of the articular surface, the salience of the *epicondylus lateralis* which is less marked than in the former subgroup, and flattening of the distal diaphysis. These subgroups would consist of *Australopithecus.*

Thus, it appears that in most Australopithecines, the *fossa olecrani* was shallow and central, leading to an almost equivalent width of the lateral and medial pillars. To explain this morphology, I decided to outline the distal extremities. Some similar studies had been done by Hultkrantz (1897) who sectioned the bones, and then by Dokládal (1977) who unfortunately only sectioned the human diaphysis. Le Floch

1 Laboratoire d'Anthropologie du Museum National d'Histoire Naturelle, Musée de l'Homme, Place du Trocadéro, 75116 Paris, France

(1978a) studied cross sections of human distal humeri in different planes and proved the opposition of the pillars in the distal humerus.

To begin with, I decided to outline bones in the transversal plane in some hominoid living primates, and applied the method to some casts of Plio-Pleistocene hominids.

Method

I used a bone support to maintain the proximal part of the bone, and a craniophore with a fixed clamp. Against this clamp was pressed the posterior flat surface of the distal humerus. Then, the bone was considered as vertical. The most salient points of the epicondyles were placed on the same horizontal line with the parallelograph (Fig. 1).

It was, thus, possible to outline the diaphysis every 2 mm from the biepicondylar width up to 20 mm. The last drawing is always set above the *fossa olecrani* (Figs. 2 and 3).

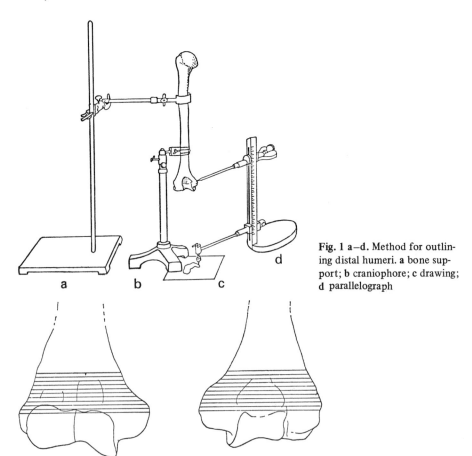

Fig. 1 a–d. Method for outlining distal humeri. a bone support; b craniophore; c drawing; d parallelograph

Fig. 2. Schema of the levels of outlines, realized on *Homo sapiens* (*left,* anterior view; *right,* posterior view)

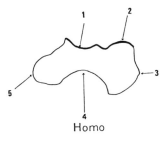

Fig. 3. Configuration of the first outline in *Homo sapiens* and *Pan: 1, trochlea humeri; 2, capitulum humeri; 3, epicondylus lateralis; 4, fossa olecrani; 5, epicondylus medialis; 6, fossa coronoidea; 7, fossa radialis; heavy line,* articular surface

Homo

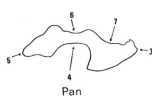

Pan

Material

In this study I only took into consideration wild, adult, male or female primates, right or left bones. Studied were 20 modern *Homo sapiens,* 40 *Pan troglodytes,* 20 *Pan paniscus,* 23 *Pongo pygmaeus* and 3 casts of Plio-Pleistocene hominids, representing each group previously described. A fiber glass cast of KNM ER 739 from East Turkana, and plaster casts of Gombore IB 7594 from Melka Kunture and AL 288.1M from Hadar were used. In this first study, neither *Gorilla* nor *Hylobates* were observed.

Comparative Anatomy

Although *Pan, Pongo* and *Homo* are related in classification, they show strong differences in distal humeral morphology.

Anterior Views (Fig. 4)

In modern Man the *capitulum humeri,* and the *trochlea humeri* are isolated by a poorly developed crest. The *epicondylus lateralis* is set almost on the same level as the *capitulum humeri.* In *Pan,* the double-trochlea pattern of the articular surface is striking: the *capitulum humeri* is isolated from the *trochlea humeri* by a salient crest. The very salient *epicondylus lateralis* projects markedly above the *capitulum humeri.* In *Pongo* the lateral crest of the *trochlea humeri* is generally salient and the double-trochlea pattern of the articular surface is present most of the time. The *capitulum humeri* is set below the laterally projected *epicondylus lateralis* as is the case in *Pan.*

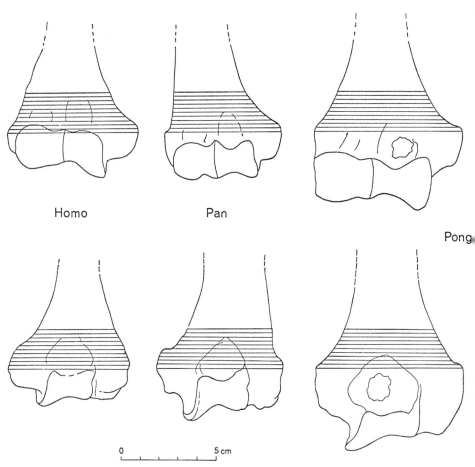

Fig. 4. Comparative anatomy of hominoid primates. *Above:* anterior views; *below:* posterior views

Posterior Views (Fig. 4)

In modern Man, the *fossa olecrani* is never limited by a strong lateral wall. The medial and lateral pillars bend gently toward the center of the fossa. In *Pan*, a strong lateral wall always limits the *fossa olecrani*, and the medial one bends smoothly toward the middle of the fossa. Most of the time, the lateral supraepicondylar crest is straight. In *Pongo*, as in *Pan*, the *fossa olecrani* is always limited by a strong lateral crest, but generally less well expressed than in *Pan*.

Lateral Views

In Man the anterior border of the diaphysis is projected anteriorly and the shaft never exhibits a strong anteroposterior flattening. In *Pan* the anteroposterior flattening of the distal shaft is marked. In *Pongo* a strong anteroposterior flattening of the distal diaphysis can be observed. It is more marked than in *Pan*.

I have to clarify that the morphology of *Pan paniscus* is roughly the same as *Pan troglodytes*. The main difference between the two species occurs in the size of the specimens.

Results

Comparison Between the Living Primates (Figs. 5 and 6)

In all pongids here studied the first outline never crosses the articular surface. The epicondyles are always set above the articular area and especially the *epicondylus lateralis* projects markedly above the *capitulum humeri*.

The *fossa olecrani* is always limited by a strong lateral wall in pongids, but never in modern Man. In this last case, both medial and lateral walls bend smoothly toward the center of the fossa.

The most striking feature is the morphology of the pillars. In modern Man, the lateral pillar is always quadrangular and the medial one triangular, which confirms Le Floch's study (1978a). However, in pongids both pillars are triangular in shape. In chimpanzees (*Pan paniscus* as well as *Pan troglodytes*) the lateral pillar presents an apex set posteriorly and the medial one is almost isosceles. In *Pongo*, the lateral pillar resembles the configuration seen in *Pan*, but the medial one is more antero-posteriorly flattened.

The flattening of the diaphysis is clearly marked in pongids: antero-posterior in *Pongo*, more oblique in *Pan*; but in modern Man the anterior salience of the anterior border gives a triangular shape to the shaft. The salience of the anterior border is displaced on the medial side in chimpanzees, probably as a result of the insertion of the *m. brachioradialis* which gives a strong flattening in the lateral area.

The lateral border of the diaphysis is, thus, almost rectilinear in chimpanzees, but in modern Man and *Pongo*, both medial and lateral sides decrease upward on the diaphysis.

Plio-Pleistocene Hominids

KNM ER 739: Discovered in 1970 above the Lower/Middle Tuff at Koobi Fora, this right humerus is remarkable in its robusticity and muscle impressions (Leakey et al. 1972, Leakey 1971). It is dated at approximately 1.3 million years (Fitch and Miller 1976) (Figs. 7 and 8). As is the case in pongids, the first outline never crosses the articular surface. Its lateral pillar is distally subquadrangular, and exhibits a marked anteroposterior flattening. The shallow *fossa olecrani* is not limited by a strong lateral crest. The weak anterior salience of the anterior border is displaced medially and the lateral supraepicondylar crest is well expressed. In total pattern, the medial and lateral borders decrease simultaneously in the first outlines; then, the medial side exhibits the strongest decrease upward. A marked anteroposterior flattening of the distal diaphysis can be observed.

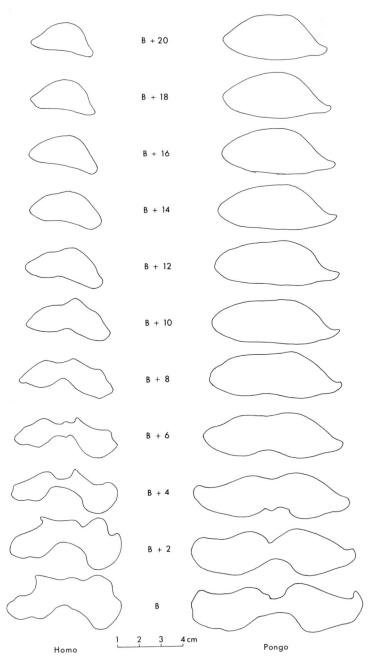

B + 20

B + 18

B + 16

B + 14

B + 12

B + 10

B + 8

B + 6

B + 4

B + 2

B

Homo   1   2   3   4 cm   Pongo

Fig. 5. Comparison of outlines between *Homo* and *Pongo* (right humeri). In all drawings the *epicondylus medialis* is used as reference. The distance separating two medial epicondyles is 2.5 cm. *B*, biepicondylar width

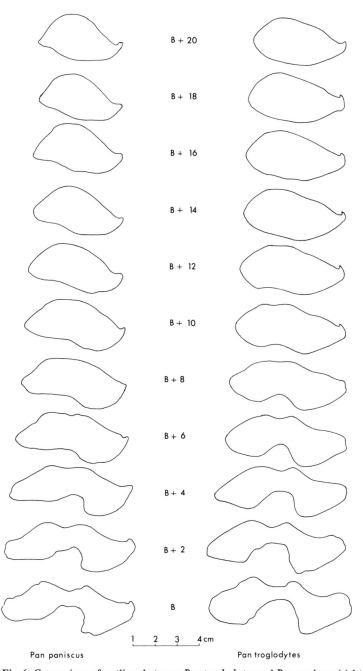

B + 20

B + 18

B + 16

B + 14

B + 12

B + 10

B + 8

B + 6

B + 4

B + 2

B

1 2 3 4 cm

Pan paniscus   Pan troglodytes

Fig. 6. Comparison of outlines between *Pan troglodytes* and *Pan paniscus* (right humeri)

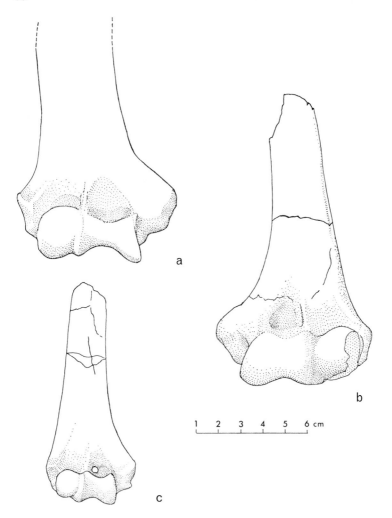

Fig. 7 a–c. Plio-Pleistocene hominids (from original specimens). a KNM ER 739 (right humerus); b Gombore IB 7594 (left humerus); c AL 288.1M (right humerus)

Gombore IB 7594: This well-preserved left distal humerus, discovered in 1976 at Melka Kunture in Ethiopia, is dated at more than 1.7 million years (Chavaillon, pers. comm.). Its general features, quite modern, permit its attribution to *Homo sp.* (Chavaillon et al. 1977; Senut 1979) (Figs. 7 and 8). As is the case in modern Man, the first outline crosses the articular surface. The anterior border of the diaphysis is well salient anteriorly and in the median area of the bone. In the supraepicondylar areas of the bone the pillars narrow into a lateral and a medial crest. The configuration of the pillars is strikingly similar to that of modern Man: the lateral one is markedly quadrangular and the medial one triangular. The sides of the diaphysis decrease upward on the same time. The deep *fossa olecrani* is not limited by a strong lateral crest as is the case in modern Man.

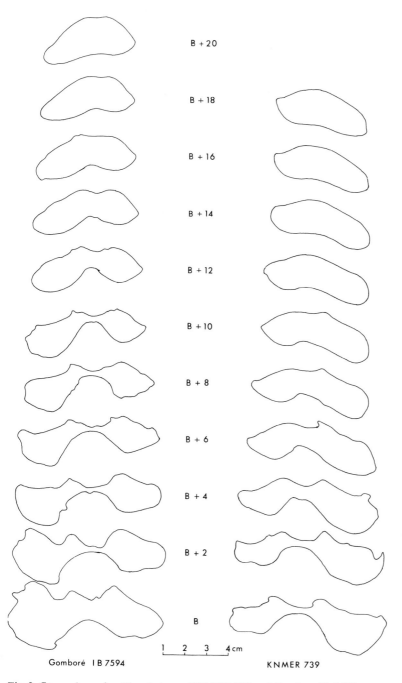

Fig. 8. Comparison of outlines between KNM ER 739 and Gombore IB 7594

AL 288 1M: Discovered in 1974 at Hadar (Ethiopia), this right complete humerus belongs to the partial skeleton attributed to *Australopithecus afarensis* (Taieb et al. 1975, Johanson et al. 1978) (Figs. 7 and 9). On the outlines, restricted to nine instead of ten (for some methodological reasons), it is clear that the first drawing does not cross the articular surface. The medial pillar is triangular, but the lateral one is not clearly a triangle. The *fossa olecrani* is not limited laterally by a strong lateral crest, but both lateral and medial pillars project above the fossa. The epicondylar border (narrow in a supraepicondylar crest) and the decreasing of the shaft are well marked in the area of the *epicondylus medialis.* The lateral border contrasts with the medial one in its rectitude which reminds one of the configuration seen in *Pan.* The distal diaphysis is flattened anteroposteriorly distally, but upward the salience of the anterior border is clear.

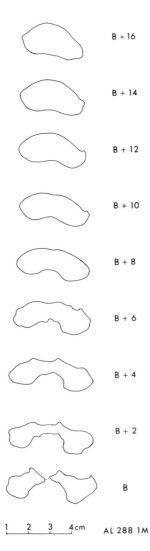

B + 16

B + 14

B + 12

B + 10

B + 8

B + 6

B + 4

B + 2

B

1 2 3 4 cm   AL 288 1M   **Fig. 9.** Outlines in AL 288.1M

Discussion

From this comparative study it appears that the distinction among the hominoids is quite possible from the distal humeri. Modern Man and pongids are different in the configuration of the pillars surrounding the *fossa olecrani*. In *Homo sapiens*, the lateral quadrangular pillar contrasts with the medial triangular one. In pongids, both of them are triangular, but in *Pongo* the flattenings are extreme. Le Floch (1978b) explains that the lateral pillar could be considered as a strength-bearing column which gives a quadrangular shape and that the medial one is mainly a muscle-bearing column which appears in its triangular configuration. It is remarkable that the nonhuman primates studied in this article represent suspensory forms of animals and we could suggest that the triangular shape of both pillars could be related to their pattern of locomotion.

Applying the method to Plio-Pleistocene hominids, two ideas have to be pointed out: first, it is possible from the outlines of distal humeri to distinguish three main groups among Plio-Pleistocene hominids, confirming a previous anatomical work (Senut 1978, 1980). Then, Gombore IB 7594, which was primitively attributed to the genus *Homo* (Chavaillon et al. 1977, Senut 1979) cannot be differentiated from a typical modern human. The other specimens share features with modern Man and some pongids (e.g., *Pan*), but Afar resembles more the chimpanzees (although it is not a pongid) in the pillar morphology.

Conclusion

This study clearly demonstrates that the outlines of bones can be useful in the differentiation among primates (in this case, some hominoid primates). It is interesting to note that they can be used in distinguishing fossil hominids, and here in the isolation of different groups among Plio-Pleistocene hominids. The function of the features observed is still difficult to assess, but I think that biomechanical researches as well as cineradiographical works on large samples of primates would permit clarification of these problems.

Although the postcranial material (and mainly the upper limb skeleton) has been slighted for many years, it appears from its description that it could be of great importance in the taxonomy of fossil Man.

Acknowledgments. I wish to thank Dr. P. Napier (London), Dr. C. Smeenk (Leiden), Dr. D. Thys van den Audenaerde (Tervuren) for permission to study the collections of primates in their care, and Professor Y. Coppens (Paris), Dr. D.C. Johanson (Cleveland), and R.E.F. Leakey (Nairobi) for providing casts of fossil hominids.

Professor A. Delmas (Paris) has to be acknowledged for his wise guidance, and special thanks are due to Dr. P. Le Floch (Paris) for his kindness in allowing me to carry out this study and for valuable discussion and criticisms.

I am also grateful to the International Afar Research Expedition, the Franco-Ethiopian Research Expedition, and the Koobi Fora Research Project.

I would like to thank the R.C.P. 292 (Centre National de la Recherche Scientifique) and the Department of Anatomy of the Faculty of Medicine for their support.

References

Broom RL, Robinson JT, Schepers GWH (1950) Sterkfontein Ape-Man *Plesianthropus*. Transvaal Mus Mem 4

Chavaillon J, Chavaillon N, Coppens Y, Senut B (1977) Présence d'Hominidé dans le site oldowayen de Gomboré IB à Melka Kunturé, Ethiopie. CR Acad Sci Paris 285D:961–963

Day MH (1976) Hominid postcranial remains from the East Rudolf succession: a review. In: Coppens Y, Howell FC, Isaac GLl, Leakey REF (eds) Earliest Man and environments in the Lake Rudolf Basin. Univ Chicago Press, Chicago, pp 507–521

Day MH (1978) Functional interpretation of the morphology of postcranial remains of early African hominids. In: Jolly CJ (ed) Early hominids of Africa. Duckworth, London, pp 311–345

Day MH, Leakey REF, Walker AC, Wood BA (1976) New hominids from East Turkana, Kenya. Am J Phys Anthropol 45:369–436

Dokládal M (1977) Variability of the cross-section shape of the shaft of the humerus and its practical significance. Folia Morphol XXV 4:343–349

Fitch FJ, Miller JA (1976) Conventional potassium-argon and argon 40/argon 39 dating of volcanic rocks from East Rudolf. In: Coppens Y, Howell FC, Isaac GLl, Leakey REF (eds) Earliest Man and environments in the Lake Rudolf Basin. Univ Chicago Press, Chicago, pp 123–147

Hultkrantz JW (1897) Das Ellenbogengelenk und seine Mechanik. Eine anatomische Studie. Fischer, Jena

Johanson DC, White TD, Coppens Y (1978) A new species of the genus *Australopithecus* (Primates: Hominidae) from the Pliocene of Eastern Africa. Kirtlandia 28:1–14

Leakey REF (1971) Further evidence of Lower Pleistocene hominids from East Rudolf, Kenya. Nature (London) 231:241–245

Leakey REF, Mungai JM, Walker AC (1972) New Australopithecines from East Rudolf, Kenya II. Am J Phys Anthropol 36:235–252

Le Floch P (1978a) Les piliers de la palette humérale. Mem Lab Anat Fac Med Paris 35:1–175

Le Floch P (1978b) Asymétrie de la palette humérale. II. Approche morphologique. Bull Soc Anat Paris 3:3–11

McHenry HM (1973) Early hominid humerus from East Rudolf, Kenya. Science 180:739–741

McHenry HM, Corruccini RS (1975) Distal humerus in hominoid evolution. Folia Primatol 23: 227–244

Oxnard CE (1973) Functional inferences from morphometrics: problems posed by uniqueness and diversity among the primates. Syst Zool 22:409–424

Oxnard CE (1975) Uniqueness and diversity in human evolution: morphometric studies of *Australopithecus*. Univ Chicago Press, Chicago

Robinson JT (1972) Early hominid posture and locomotion. Univ Chicago Press, Chicago

Senut B (1978) Contribution à l'étude de l'humérus et de ses articulations chez les Hominidés du Plio-pléistocène. Thèse Doc 3ème Cycle Paléontologie Hum Univ Pierre et Marie Curie, Paris

Senut B (1979) Comparaison des Hominidés de Gomboré IB et de Kanapoi: deux pièces du genre *Homo?* Bull Mem Soc Anthropol Paris 6:111–117

Senut B (1980) New data on the humerus and its joints in plio-pleistocene Hominids. Coll Anthropol 4:1

Straus WL (1948) The humerus of *Paranthropus robustus*. Am J Phys Anthropol 6:285–311

Taieb M, Johanson DC, Coppens Y (1975) Expédition internationale de l'Afar, Ethiopie (3ème campagne 1974). Découverte d'Hominidés Plio-Pléistocènes à Hadar. CR Acad Sci Paris 281D: 1297–1300

Structural-Functional Relationships Between Masticatory Biomechanics, Skeletal Biology and Craniofacial Development in Primates

O.J. Oyen and D.H. Enlow [1]

It has been repeatedly demonstrated that overall skeletal growth is influenced in general by genetics, nutrition, mechanical stimulus, and by endocrine function. Some of the specific control factors that govern bone growth at the cellular level have also been characterized (Bourne, 1956–1979, provides an extensive review). The validity of the concept that during growth a bone adapts to functional forces that act upon it (Wolff's Law) has been well established. Still, numerous questions persist about the mechanical and biochemical events that provide for site-specific control of localized growth processes. For example, it remains to be determined how it is that parts of the skull such as the supraorbital and nuchal regions have the capacity for highly localized bone growth patterns that are characteristic for these regions as compared to others.

Part of our research over the past few years has focused on the identification and elucidation of surface characteristics and patterns of trabecular organization in cross-sectional collections of extant and fossil primates (Oyen 1974, 1977, Oyen et al. 1979a,b, Oyen and Rice 1980, Oyen et al. 1981). While most of the work carried out thus far has focused on the ontogenesis and phylogenesis of the supraorbital region, our attentions have recently been broadened to include other parts of the skull. What follows is a brief report in which localized bone growth patterns in the supra-orbital and the nuchal regions are compared with each other and with growth-related changes in the masticatory system in several different primate species. In this preliminary analysis, data obtained in earlier studies are combined with additional information about the histological structure and development of the nuchal region. An attempt is made to show that, upon the basis of these combined data, such seemingly unrelated parts of the skull as the browridge and the nuchal region are unified through structural, functional and developmental links with another part of the skull, the masticatory system.

1 School of Dentistry, Case Western Reserve University, 2123 Abington Road, Cleveland, Ohio 44106, USA

Materials and Methods

The specimen samples and procedures used in this study have been, for the most part, described in detail elsewhere (see Oyen et al. 1979a,b, 1981, Oyen and Rice 1980). Briefly, they are as follows: subadult and adult skulls from extant species including chimpanzees (*Pan* sp.), gorillas *(Gorilla gorilla)*, orangutans *(Pongo pygmasus)*, gibbons *(Hylobates lar)*, olive baboons *(Papio cynocephalus anubis)*, gelada baboons *(Theropithecus gelada)*, and macaque monkeys *(Macaca mulatta)* were studied. Fossil hominid materials analyzed included the Pech de l'Aze infant and the La Quina 5, Broken Hill, and La Chapell-aux-Saints skulls. Also considered were the East African fossils crania KNM-ER 406 *(Australopithecus robustus)*, KNM-ER 1470 *(Homo habilis)*, KNM-ER 1813 *(Australopithecus africanus)*, KNM-ER 3733 *(Homo erectus)*, the Bodo d'Ar specimen and Laetolil hominid 21 *(Australopithecus afarensis)*.

The topographic features of the cortical bone in the supraorbital and the nuchal regions were noted in each of the skulls, particular attention being paid to the porosity and texture of the bone in these areas. Samples of bone were taken from these regions in selected subadult and adult baboons and chimpanzees. The bone sections were prepared and analyzed using standard histological procedures. For purposes of comparison, cortical bone posterior to the supraorbital sulcus and overlying the neurocranium was also examined.

Results

In approximately two-thirds of the skulls the cortical surface in the supraorbital region was marked by a vermiculate or worm-track appearance (Fig. 1).

Histological examination of bone section showed this pattern and much of the lamina externa in the supraorbital regions to be formed by unconsolidated or partially consolidated deposits of fine cancellous bone. Variations in the distribution of different surface patterns were found to be a reflection of the degree to which the layers of porous, fine cancellous bone had been converted by cancellous compaction into typical compact-appearing bone of the lamina externa. In contrast, cortical bone of the neurocranium, 4—5 cm posterior to the supraorbital sulcus, was more homogeneous compact bone consisting of lamellar and parallel fibered primary vascular bone, with secondary (Haversian) deposits present in the more mature skulls. More detailed descriptions of the histology of these tissues are provided in Oyen et al. (1979a,b, 1981) and Oyen and Rice (1980).

When the nuchal region was briefly examined in each of the species included in this preliminary study, we found that cyclic deposits of fine cancellous bone were also present in the nuchal region more or less at the same time they occur in the brow region.

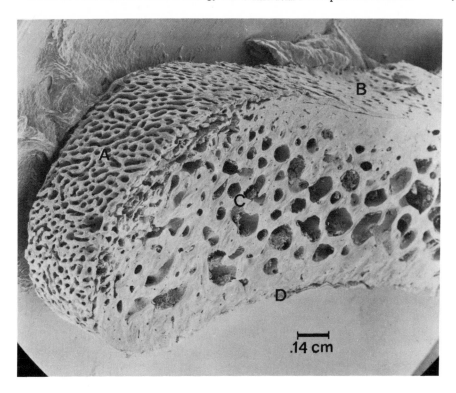

Fig. 1. Scanning electron micrograph of bone section from supraorbital region in a subadult chimpanzee (7 X). *A* Fine cancellous bone with characteristic vermiculate cortical surface pattern along anterior supraorbital margin. *B* Semiconsolidated fine cancellous bone along posterior margin and sulcus. *C* Coarse cancellous bone of diploe visible in cut section. *D* Compact primary bone of orbital roof

Discussion

In our first studies of olive baboons (Oyen 1974, Oyen et al. 1979b), it was noted that structural-functional relationships seemed to exist between the presence of fine cancellous bone in the brow region and growth-related changes in the masticatory system. Further studies (Oyen 1977, Oyen et al. 1979a, 1981, Oyen and Rice 1980, work in progress) have provided evidence that similar though not necessarily identical relationships seem to exist between the development of the supraorbital region and the masticatory system in other primates including macaque monkeys, gelada baboons, all of the apes and in hominids.

On the basis of these studies we have suggested that browridge development in these primate species occurs in response to growth-related changes in the forces generated during chewing. Given the presence of epidosic deposits of rapid-growing bone tissue in the nuchal region which seem to coincide with bone growth activity in the supraorbital region, we would now like to expand our model of skull growth.

In this expanded model, it is suggested that growth and development of the supraorbital and nuchal regions are linked with each other through their common respon-
siveness to growth-related changes in the masticatory system. Growth in our model
occurs more or less in the following fashion.

The neonate skull represents a shape well adapted for the immediate dietary-masti-
catory needs of a suckling infant. The nutritional needs and masticatory functions of
the infant change with the development and eruption of teeth and the addition of
chewed foods to the diet. To accomodate the increased spatial needs and force patterns associated with mastication, the skull must also respond by means of growth-
related remodeling.

Additions to the tooth row that contribute to the formation of a masticatory load
arm that is longer than the lever arm generally lead to an increase in the overall forces
generated during chewing. In the earlier model, it was maintained that the episodic
development of the supraorbital region represented a positive response of the frontal
bone to such increases in masticatory force. In this expanded model, it is suggested
that the effects of tooth eruption on the development of the skull extend beyond
a simple elongation of masticatory load arm length.

In Fig. 2 the outline of an infant baboon skull is superimposed on an adult skull
in order to illustrate the effects of progressive enlargement of the face on the balance
of the skull on the vertebral column. Each skull rests on the column and functions
as a first order lever. The fulcrum of each skull is provided by the occipital condyles.
For purposes of this analysis, the load arm of the skull is determined by the length of
a horizontal plane extending perpendicularly from the most anterior point on the
skull back to a vertical plane perpendicular to the occlusal plane of the occipital condyles. Lever arm length of the skull extends from the vertical plane of the occipital
condyles to the most posterior point of attachment of the nuchal muscles.

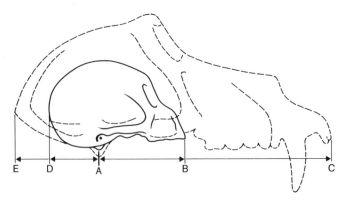

Fig. 2. Growth-related differences in cranial lever arm and load arm lengths. *Solid line*, edentolous
neonate baboon skull. *Dotted line*, Adult male baboon skull. Large increases in load arm lengths
(*lines A—B* and *A—C*) between infancy and adulthood are countered in part by increases in lever
arm lengths (*lines A—D* and *A—E*) and increases in the musculature which is in the nuchal region.
The differences between load arm length and lever arm length are exaggerated in the genus *Papio;*
however, large changes in load arm length as well as load size are characteristic of facial and dental
development in all mammals including primates

As teeth develop and enlarge, their increasing weight contributes to a disequilibrium in the balance of the skull because all teeth in primates develop anterior to the occipital condyles. Moreover, as the tooth rows elongate, the additional weight of the teeth is displaced more and more from the cranial fulcrum, thereby contributing even more so to an anterior tilt of the skull. Added to these factors are increases of the mass of the chewing muscles associated with greater masticatory forces.

Developmental changes in the occipital region of the skull are essential to compensate for the anterior disequilibrium of the skull that has been described. While neurocranial expansion and increased thickness of the braincase posterior to the occipital condyles may provide some counterweighting, the primary mechanism for countering growth-related increases in the mass of the face is provided by the nuchal musculature. Thus, as the face grows and the masticatory system develops, the muscles which are attached in the nuchal region, i.e., mm. trapezius, semispinalis capitis, and rectus capitis major and minor, respond and undergo compensatory growth which provides the necessary counterbalancing forces. In association with the increased muscular activity and growth, the nuchal lines and crests become more pronounced.

In this model, it is suggested that because changes in the masticatory complex occur in an episodic fashion that reflect tooth development and eruption, morphogenesis of the supraorbital and nuchal regions also occurs in an episodic, rather than a continuous manner. In this model, patterns of bone growth can be linked with structural, functional, and developmental aspects of masticatory biomechanics in providing a testable hypothesis that reflects the unity of seemingly unrelated parts of the skull. It must be noted, however, that this is a preliminary report whose results have yet to be experimentally verified. More detailed studies, which quantitatively assess the relationships between bone growth patterns in the nuchal region, as well as other parts of the skull with masticatory growth and function, are called for.

References

Bourne GH (ed) (1956/79) The biochemistry and physiology of bone, vols I–IV. Academic Press, London New York

Oyen OJ (1974) The baboon face. Unpubl. Doct Diss, Univ Minnesota. Univ Microfilms, Ann Arbor, Mich

Oyen OJ (1977) Appositional bone growth in Neanderthalensis: An exercise in comparative osteology. Am J Phys Anthropol 47:153–154

Oyen OJ, Rice RW (1980) Supraorbital development in chimpanzees, macaques and baboons. J Med Primatol 9:161–168

Oyen OJ, Rice RW, Cannon MS (1979a) Browridge structure and function in extant primates and Neanderthals. Am J Phys Anthropol 51:83–96

Oyen OJ, Walker AC, Rice RW (1979b) Craniofacial growth in olive baboons *(Papio cynocephalus anubis)*: Browridge formation. Growth 43:174–187

Oyen OJ, Rice RW, Enlow DH (1981) Cortical surface patterns in human and non-human primates. Am J Phys Anthropol 54:415–419

Comparison of Morphological Factors in the Cranial Variation of the Great Apes and Man

B. Jacobshagen [1]

For about one century there has been a continuous discussion among morphologists concerning the causes of the typical differences in the facial skeleton of the hominoid species. The complementary views primarily referred to concern the supraorbital ridging or its lack. Some authors attribute this characteristic exclusively to the spatial relationships between brain and orbits (Moss and Young 1960). Others suppose an influence of the mechanical stresses due to mastication (discussed by Biegert 1957, Ehara and Seiler 1970, Vogel 1966). The genetically determined spatial conditions might be insufficient for function without stress-induced growth processes, which improve the capacity for mechanical stress (similar to the growth of cranial crests).

Against the background of this controversy, the purpose of this study is to find out basic correlation patterns of craniometric landmarks and their spatial distributions.

Material and Methods

The investigations were based on five samples of adult hominoid skulls: *Pan, Pongo* (males only), *Homo* (mixed, mostly males) and *Gorilla* (males and females). Sample size was 25 to 30. For the determination of variables a new biostereometric technique was applied. The stereophotogrammetric system after Kellner can be used with a pair of normal 35 mm cameras, insofar as the optical distortions of the lenses are precisely documented for later compensation. Of further importance is the use of a three-dimensional reference-point system surrounding the object. The whole configuration used is shown in Fig. 1. The additional slide projector (left) enables better identification of unstructured surfaces using a random-pattern slide which forms up corresponding marks necessary for surface identification during measurement (Fig. 2). An example of the resulting stereoscopic pairs is given in Fig. 3. Relatively strong parallaxes are apparent due to the relation of inter-camera distance and object distance of 30 cm: 120 cm = 1:4. The purpose of enlarged parallaxes is an improved depth resolution of some four to five times as compared with human interocular distance.

The measurements of anatomical landmarks were taken using an analytical stereo-comparator (Fig. 4) at the Department of Biophotogrammetry of Göttingen University.

1 Anthropologisches Institut der Universität, Von-Melle-Park 10, 2000 Hamburg 13, FRG

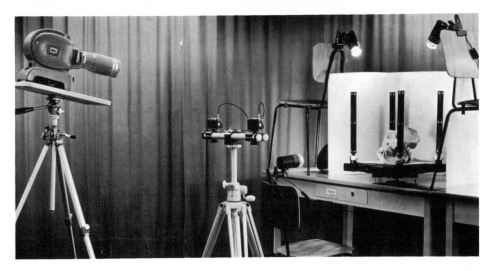

Fig. 1. Configuration for stereophotogrammetry. Cameras *(middle)*, reference-point system with skull *(right)* and slide projector *(left)*

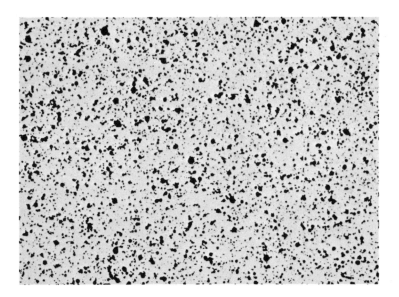

Fig. 2. Random pattern for contouring otherwise unstructured surfaces

This machinery was linked with a computer which filed 3-D coordinates whenever a foot switch was pressed. In this way the measurement of surface marks was very effective. It took some 90 min for filing the coordinates of about 300 anatomical land-marks out of six or seven stereoscopic pairs for each skull. This included some tech-nical adjustments, especially the measurements of the reference marks. Using this technique it is even worthwhile to document contours by arrays of marks, most of

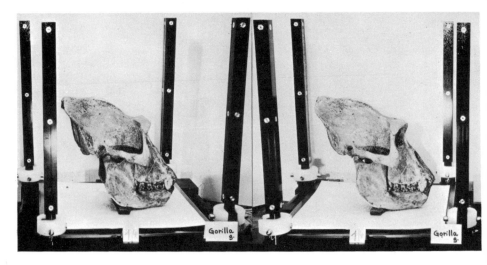

Fig. 3. Example of a stereoscopic pair

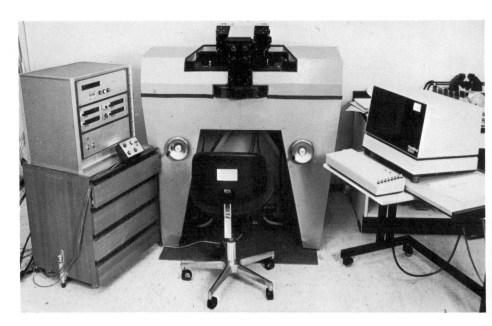

Fig. 4. Analytical stereocomparator (Dpt. of Biophotogrammetry of Göttingen University)

them necessarily undefined. This allows better object reconstruction and contains more biological information than conventional distances (Jacobshagen 1980; Lestrel 1974, 1975, Moyers and Bookstein 1979). As an example, Fig. 5 demonstrates the density of measurements of a human skull.

The comparability of the data was achieved after converting the coordinates of each skull to a common coordinate system based on a slightly modified Frankfurt-plane (Fig. 6). For this study, only reproducible landmarks were evalutated, The x, y, and z coordinates of up to 58 landmarks were fed into factor analyses using varimax rotation (BMD 08), separately for each species. Taking into account the reversive m/n-relation for these computations as recommended by Weber (1974), some test analyses were made using a "correct" number of variables (m = n/3). Comparing the results, no basic differences were found in the factor patterns. This suggests that no essential distortion occurred.

Results

It is interesting to note that the majority of factors representing the greatest amount of variance can be attributed to certain coordinate axes. So the factor patterns of each component are demonstrated separately using slightly generalized "maps" of the facial skeleton.

Regarding the factor patterns for the sagittal (y-) component (Fig. 7), there is only one factor in *Pan, Gorilla* and Man, but two factors in *Pongo*. In all samples except male *Gorillas* this one is the first factor which is associated with general size [2]. The figures do not show the always separate factors describing the variation of the cranial base.

Looking at the factors describing correlation structures along the lateral (x-) axis (Fig. 8) there is a general "facial x factor" to be found in *Pan, Gorilla* and *Pongo*. *Gorilla* shows another type of factor covering lateral areas; an "upper-face factor" in males and females and a "lower-face factor" in females only. Quite different from these patterns is the factor structure in Man. Here are two sidewise complementary factors in the upper face. Additionally, a part of the mandible seems to be influenced symmetrically by the left-side factor.

Regarding the vertical (z-) component, three factors show surprisingly constant regional patterns (Fig. 9). The most important one always is a "jaw factor". Contrary to the great apes, this one does not include the frontal maxillary and nasal area in Man. The next important z factor influences the supraorbital region, more or less including the upper part of the orbital margin, especially in *Pongo*. Another factor represents a correlation pattern limited to the zygomatic arch and zygomatic bone, concentrated on its frontal and lower part. In *Pan* there is a separate factor for the lateral part of the zygomatic arch.

Generally, there is correspondence to a great extent both in the factor patterns and the rank order of factors within the spatial components.

---

2   There are a few additional loadings of x and/or z coordinates which are not illustrated here

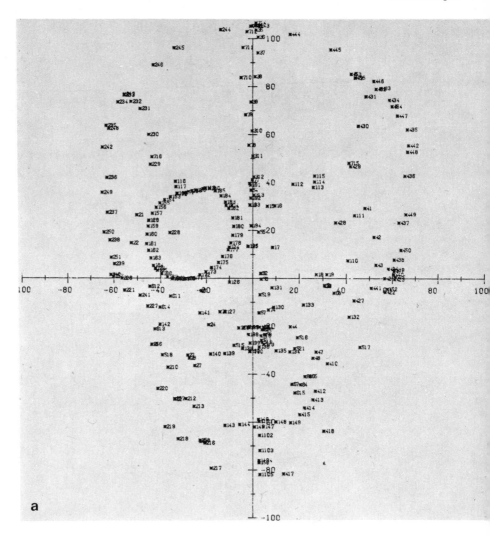

a

Fig. 5. a Frontal and b lateral projection of three-dimensional data (human skull)

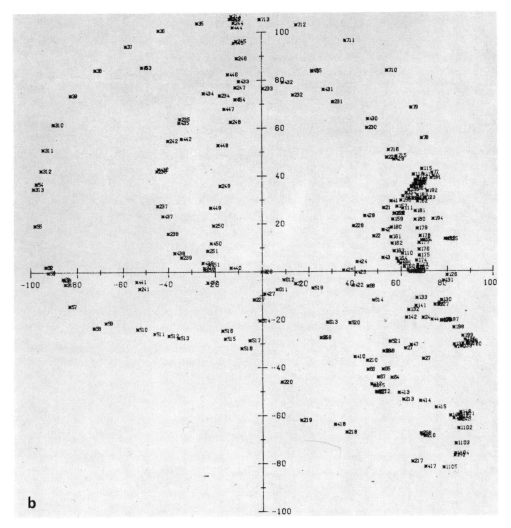

Fig. 5. b

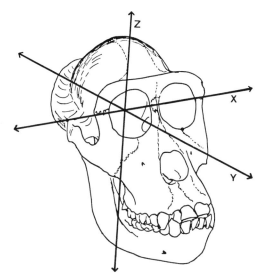

Fig. 6. Coordinate system; x-y-plane defined by the auricular points and orbitale inferior *(left side);* position of origin is in the *middle* of the auricular points

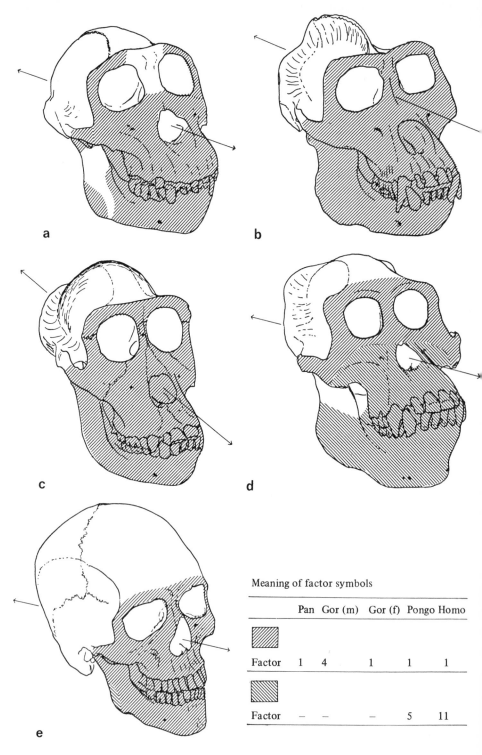

Meaning of factor symbols

	Pan	Gor (m)	Gor (f)	Pongo	Homo
Factor	1	4	1	1	1
Factor	–	–	–	5	11

Fig. 7 a–e. Factor patterns of y-component. **a** *Pan* (males); **b** *Gorilla* (males); **c** *Gorilla* (females); **d** *Pongo* (males); **e** *Homo* (mixed, mostly males)

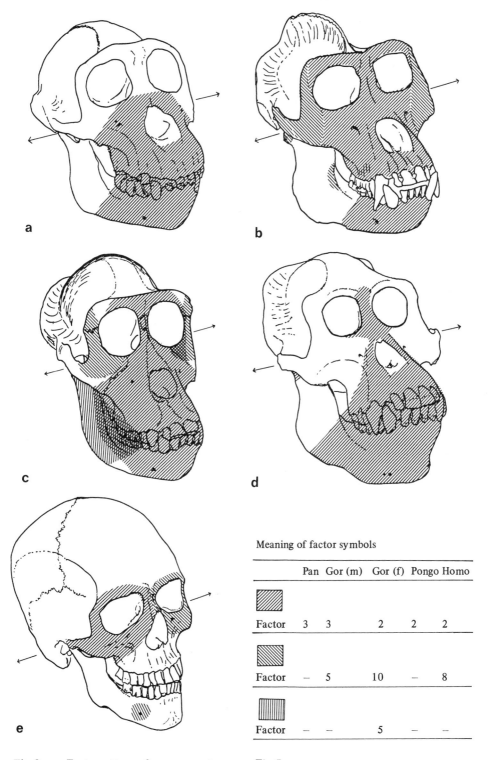

Meaning of factor symbols

	Pan	Gor (m)	Gor (f)	Pongo	Homo
Factor	3	3	2	2	2
Factor	–	5	10	–	8
Factor	–	–	5	–	–

Fig. 8 a–e. Factor patterns of x-component. a–e see Fig. 7

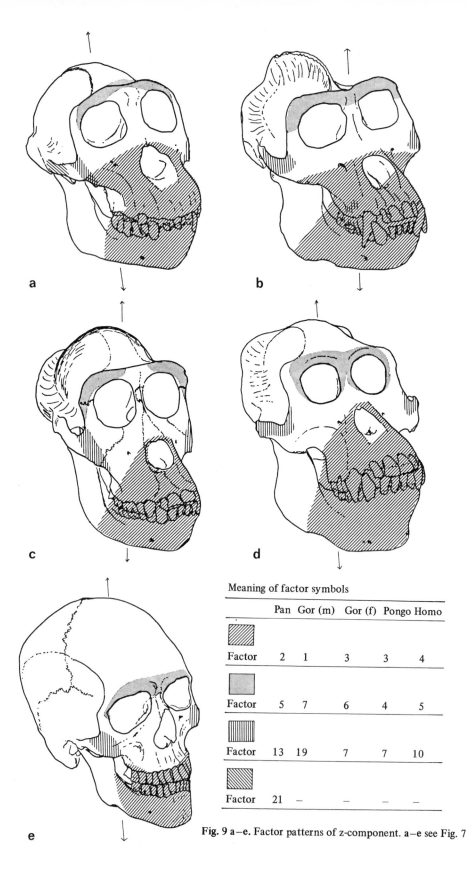

	Pan	Gor (m)	Gor (f)	Pongo	Homo
Factor	2	1	3	3	4
Factor	5	7	6	4	5
Factor	13	19	7	7	10
Factor	21	–	–	–	–

Meaning of factor symbols

Fig. 9 a–e. Factor patterns of z-component. a–e see Fig. 7

Discussion and Conclusions

Although the number of variables posed problems statistically to factor analysis, the conformity of the results suggests a sufficient degree of reliability. Looking at the factor patterns of the three axes, the following conclusions may be drawn.

The facial y factors show variation in the position of the facial landmarks as a whole relative to the origin of the coordinate system (situated in the middle of the two auricular points). That might be a result of allometric growth in the sense of allomorphosis after Giles (1956). The two distinct factors for upper and lower face in *Pongo* reflect a different growth pattern and might be an expression of the taxonomic differences between orangutan and African apes.

The x-component factors are of different types. The "facial x factor" of *Pan*, *Gorilla* and *Pongo* is not related to breadth variation. Certainly, this one is due to random variation in the shift or bending of the median-sagittal plane [3] between neurocranium and splanchnocranium. The other type of x factor is represented by the additional factors in *Gorilla* which are based on breadth variation. This is indicated by opposite signs of factor loadings when comparing bilaterally situated landmarks. The two asymmetric x factors in Man are closer to the first type than to breadth factors. In a separate factor analysis, using linear distances instead of coordinates, an asymmetry was found for (mono-) zygomatic breadth only. These measurements were derived from the basion position on the x-axis. The lack of correlations between left and right side of the human facial skeleton might be due to brain asymmetry, similar to handedness.

Contrasting with the results concerning the x-component correlations, z-component factors show a high degree of uniformity among pongids and Man. This is primarily of taxonomic significance. The structure of growth patterns is very similar. The "jaw-factor" is a result of variation in the position of jaws relative to the braincase. The "supraorbital factor" is probably due to orbital height variation. The third type of z factor is likely to be explained by the assumption of functional stress. The distribution patterns of factor loadings coincide with the attachment of the masseter muscle.

The latter one is the only factor which is undoubtedly related to mastication. The others are most probably the effects of general growth patterns. So there is only little evidence for the influence of functional stress on the variation of anatomical landmarks in the facial skeleton.

Acknowledgments. For the possibilities of data collection I am grateful to Dr. P. Napier (British Museum of Natural History/London), Mrs and Mr Powell-Cotton and Mr Barton (Powell-Cotton-Museum/Birchington, Kent) and Dr. C. Smeenk (Museum of Natural History/Leiden, Netherland) as well as Dr. M. Schultz (Anatomical Institute/Göttingen, FRG). I thank Prof. Dr. H. Kellner and Dipl. Biol. H. Zeltner for design and introduction of the photogrammetric system. I would also like to acknowledge my teacher, Prof. Dr. R. Knußmann, for his advice on questions of craniometrics and statistics. Sincere thanks go to Math. TA U. Holst for a great deal of computational work.

---

3 Biologically; as defined by landmarks and *not* in the sense of the coordinate system used here where no exact coincidence of y-z-plane and plane of facial symmetry exists

References

Biegert J (1957) Der Formwandel des Primatenschädels. Morphol Jahrb 98:77–199

Ehara A, Seiler R (1970) Die Strukturen der Überaugenregion bei den Primaten, Deutungen und Definitionen. Z Morphol Anthropol 62 (1):1–29

Jacobshagen B (1980) Grenzen konventioneller Techniken und Möglichkeiten alternativer Ansätze in der Anthropometrie. Mit einem Beispiel für den Einsatz der Biophotogrammetrie in der Schädelmeßtechnik. Z Morphol Anthropol 71 (3):306–321

Lestrel PE (1974) Some problems in the assessment of morphological size and shape differences. Yearb Phys Anthropol 18:140–162

Lestrel PE (1975) Fourier-analysis of size and shape of the human cranium: A longitudinal study from four to eighteen years of age. Unpubl. Ph D Diss, UCLA

Moss ML, Young RW (1960) A functional approach to craniology. Am J Phys Anthropol 18: 281–292

Moyers RE, Bookstein FL (1979) The inappropiateness of conventional cephalometrics. Am J Orthod 75 (6):599–617

Vogel C (1966) Morphologische Studien am Gesichtsschädel catarrhiner Primaten. Bibl Primatol, Fasc 4. S Karger, Basel

Enamel Prism Patterns of European Hominoids – and Their Phylogenetical Aspects

N.I. Xirotiris[1] and W. Henke[2]

Everybody concerned with questions of taxonomy and phylogeny knows that a large part of information used to classify fossil vertebrates is derived from teeth. This comes from the reasoning that teeth are the best mineralized portions of the skeleton and thus usually also the best preserved remains. The best preserved portion of teeth is again the most highly mineralized – the enamel. That the enamel shows a so-called prism pattern, which differs markedly within mammals and also within the primates, is well known since Carter (1922) and Regan (1930) published articles concerning the variability of enamel prism patterns. These were for the first time described by Tomes in 1848. An intensive investigation of enamel microanatomy and histochemistry was undertaken, besides others, by the anatomist Shobusawa (1952) who pointed out that these structures could possibly be good markers for taxonomic purposes within the primates. Meaningful publications were undertaken by Boyde (1965a,b, 1969, 1971), Gustafson and Gustafson (1967), Helmcke (1955, 1967), Osborn (1970), and other authors, which provided information about the way in which enamel structures are formed. For some years primatologists have been interested in this particular field in order to differentiate among fossil primates, whose remains are mostly represented by teeth and jaw fragments (as is the case in most of the Miocene hominoids). This group is of special interest as it is expected to have given rise to the hominoid lineage and lineages leading to the extant great apes. The theory of differentiation which has received widest support among palaeontologists is based on *Ramapithecus,* a Middle Miocene hominid, which is considered to be the phyletic ancestor of *Australopithecus* (Simons 1961, 1977, Simons and Pilbeam 1965; Early Divergence Hypothesis A). The other hypothesis and more recent interpretation is based on independent middle Miocene hominoid lineages which are distinguished from each other primarily on the basis of thickness of their molar enamel caps (see Pilbeam et al. 1977; Early Divergence Hypothesis B). Finally there is a so-called Late Divergence Hypothesis, presented by Greenfield, which states that the lineage leading to *Homo* did not become independent from the lineages to the extant great apes until the Middle to Late Miocene. The latter would mean that Middle Miocene *Sivapithecus* (incl. *Ramapithecus*), rather than Early Miocene *Dryopithecus,* is a last common acestor to *Pongo, Pan, Gorilla* and *Homo* (see Greenfield 1980, Henke 1981).

1 Institut für Anthropologie und Humangenetik für Biologen, Universität Frankfurt, Siesmeyer-straße 70, 6000 Frankfurt, FRG
2 Anthropologisches Institut, Johannes Gutenberg-Universität, Saarstraße 21, 6500 Mainz, FRG

The main focus of this paper is the following: Is there any difference between the enamel prism patterns of European hominoids and extant great apes as well as *Homo,* which can give us an idea of the separation of the recent groups and their ancestry?

By establishing SEM-laboratories in many universities, a good basis for intensification of studies on ultrastructures was made possible, and thus the use of microstructural features for taxonomic purposes. Recently Gantt et al. (1977), and Vbra and Grine (1978) focused on the possibility of distinguishing the hominid and the pongid line within the Hominoidea on the basis of enamel prism structures (see also Xirotiris et al. 1979, Henke and Rothe 1980, Protsch, in press). Concerning the problem of separation of the hominid lineage Gantt et al. (1977) suggested that *Ramapithecus* shares a so-called keyhole pattern with *Homo sapiens* which differs distinctively from the so-called hexagonal or circular pattern of pongids. While Gantt and Cring (1979) seem to have confirmed this result, namely that *Ramapithecus* separated from the pongids and shows similarities to *Homo sapiens,* Vbra and Grine (1978) came to another conclusion. They found in the Hominoidea (*Australopithecus, Homo,* and extant pongids) the same arrangement – a keyhole pattern. In other words, gross prism morphology contains according to their results no information on phylogenetic relationships of the hominoid species. In a pre-congress of the VIIIth International Primatological Society Gantt negated his former results, namely that *Ramapithecus* resembles recent *Homo* in his prism shape. Following his newly established typology (see abstracts of Anthropologica Contemporanea, Firenze 1980), *Ramapithecus* shows a so-called IIIA pattern which is typical for extant pongids *(Pongo),* while the IIIB type is typical for recent man. It means that *Ramapithecus* does not share the same pattern with *Homo.* In regard to the above given "divergence hypothesis", this can be interpreted as *Ramapithecus* being not yet on the lineage to *Homo* but possibly a last common ancestor, as Greenfield (1980) suggested.

Material and Methods

The dental material under study is listed in Table 1.

Table 1. List of investigated fossil material of Miocene hominoids

---

(1) *Dryopithecus fontani* (formerly designated as *D. suevicus*)
Site: Melchingen. Material: upper and lower teeth; permanent teeth and deciduum

(2) *Dryopithecus fontani* (formerly designated as *D. suevicus*)
Site: Trochtelfingen. Material: lower left molar

(3) *Anthropodus brancoi* (Schlosser 1901) – after Hürzeler
This fossil belongs to the genus *Pliopithecus* (?)
Other names:. *Neopithecus brancoli* (Abel 1902)
Dryopithecus rhenanus (Remane 1921)
Dryopithecus brancoi (Abel 1931)
Site: Salmendingen. Material: left M_3

(4) *Ouranopithecus macedoniensis* (de Bonis and Melentis 1976)
(formerly *Dryopithecus macedoniensis*)
Site: Bathylakkos, Greece. Material: Ind No. RPL 55

(5) *Rudapithecus hungaricus* (Kretzoi 1975) – by some others described as *Ramapithecus hungaricus*
Site: Rudabanya/Hungary. Material: Ind No. 46

The method applied is that described by Gantt and Cring (1979) The preparation method for *Ouranopithecus macedoniensis* was the following. The enamel surface was treated with acid (H_3PO_4) for 60 s. All others were prepared by slight polishing with a drill. The results of the REM-study are given in the Figs. 1, 2, and 4.

Dryopithecus fontani from Lechingen (Southern Germany)

The tooth investigated is an upper molar (see Fig. 1, top). The shape of the prisms can be described as circular to polygonal and does resemble those patterns which are described as pongid-like by Gantt et al. (1977). There are, however, slight differences, but one can be sure that this pattern, which is shown in Fig. 1b in a higher magnification, in no way has affinities to a recent *Homo*-keyhole pattern. It may be that this pattern is unique because the grade of resemblance to other shapes, described in the previous literature, is really small. Figure 1c represents also *Dryopithecus fontani* and we can diagnose, especially in higher magnification (Fig. 1d) of the left molar, that the pattern is hexagonal. This means that there are affinities to recent pongids, but — when looking from the left side — there are also slight affinities to a hexagonal shape. This is also the case in the pattern of the upper left deciduous tooth from Melchingen (Fig. 1e,f).

Those enamel prism patterns that are given in Fig. 2a,b are from a left lower molar of *Dryopithecus fontani* from Trochtelfingen. Though it is very hard to decide whether this pattern shows more affinities to a hexagonal or circular pattern, it seems that there are few affinities to a keyhole pattern and they are in some cases obvious.

Summarizing the results for the *Drypithecus fontani* teeth from Melchingen and Trochtelfingen we have to confess that it is very difficult to find a uniform description of patterns. One has to be very cautious in the determination of the patterns since there is no standardized method for comparison worked out yet. We do not know very much about the intraindividual variability of these patterns on one tooth as well as on different teeth of one individual. The only result we can present here is that one can find on a small section a number of different shapes and that it seems to be necessary to undertake a statistical approach in the determination of the most common pattern.

Anthropodus or Neopithecus brancoi

While Hürzeler (cited in Simons 1972, Simons and Pilbeam 1965) suggested this fossil belongs to *Pliopithecus*, Kretzoi (1976) stated that this form can not belong to the genus *Pliopithecus* if *Anapithecus* is a subgenus of this possible forerunner of the hylobatids (see Andrews 1978). The shape of the enamel structure (Fig. 2c,d) of the left M_3 from Salmendingen in Swabia is polygonal (hexagonal), and shows strong affinities to the extant pongids as shown by Gantt et al. (1977). This shape differs markedly from a keyhole pattern of *Homo*. A differential diagnosis between pongids and hylobatids can unfortunately not be made. This applies also to comparison of the *Dendropithecus*-patterns which would be very informative.

Rudapithecus hungaricus

The teeth of *R. hungaricus* resemble in their enamel prism pattern the shape of pongids and have almost no affinities to a keyhole pattern which is diagnostic for *Homo*. While the gross morphology of this species has many affinities to the genus *Homo* (not to *Australopithecus!*) as Kretzoi (1975, 1976) showed, we think that the affinities of the prisms of *R. hungaricus* with the dryopithecids on the one hand and the pongids on the other does not necessarily mean that *Rudapithecus* is on the lineage to the pongids. In our opinion, another interpretation is that this Ramapithecine shares in his features the patterns of Miocene hominoids and that the separation took place in later times. This would support the suggestion of a late divergence of hominids and pongids. We have to be careful however, because it is based only on the examination of one tooth from abundant material (see also Henke 1981).

Ouranopithecus (Dryopithecus) macedoniensis

One of the most interesting forms of Miocene hominoids described in the last years is *Ouranopithecus macedoniensis* (de Bonis and Melentis 1976, de Bonis et al. 1974) (Fig. 3). The question whether this fossil belongs to *Sivapithecus* (Pilbeam et al. 1977) or *Dryopithecus* (Olshan 1979) or whether it goes together with the genus *Gigantopithecus*, can not be solved presently. While *Gigantopithecus* is described as having an arcade prism pattern, *Ouranopithecus* seems to have more affinities to the genus *Sivapithecus* because of his more circular enamel prism pattern (Fig. 4).

Conclusions

If we look at the above results of SEM-microscopy we can conclude that the most common pattern seen in the European Miocene/Late Pliocene hominoids under investigation are mainly pongid-like or indecisive. The research on enamel prisms is still in an early period and we should be extremely cautious in interpreting the results taxonomically. We are not yet at the stage to build phylogenetic trees. There is a good hope however that, if we get more information about the variability of the patterns (intraindividual, intra- and interspecific), these structures can serve for taxonomic purposes. In this way we regard our work as a necessary first step only. The conclusion that the Miocene hominoids do not differ from one another in the prisms can be refuted if we look at the figures given even though it is not yet time to define them. We can say, however, that they nearly always differ from the *Homo*-keyhole pattern. The result signifies that the European Ramapithecines and possible Sivapithecines *(Ouranopithecus)* do not differ markedly from the Dryopithecines. ·

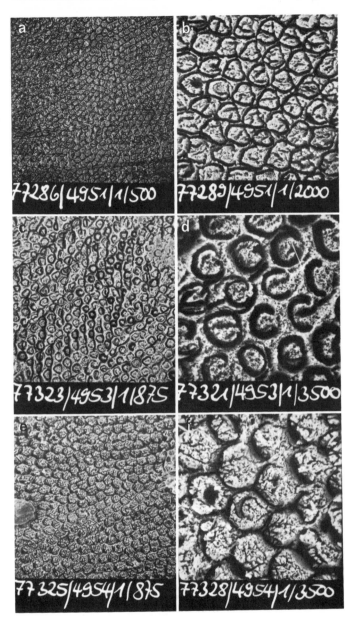

Fig. 1 a–f. a, b Enamel prism pattern of *Dryopithecus fontani* Lartet (formerly *Dryopithecus suevicus* Koken). Upper right molar from Melchingen (FRG). Magnification is always written on the right side of the subscription referring to an REM-figure 10 x 10 cm. **c, d** *Dryopithecus fontani* Lartet (formerly *Dryopithecus suevicus* Koken). Upper left deciduous molar from Melchingen (FRG). **e, f** *Dryopithecus fontani* Lartet (formerly *Dryopithecus suevicus* Koken). Lower right molar from Melchingen (FRG)

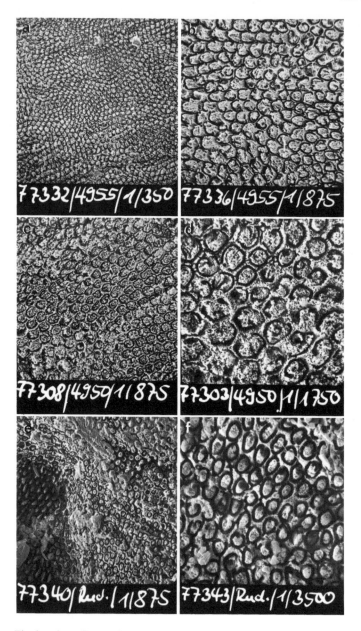

Fig. 2 a–f. a,b *Dryopithecus suevicus* Lartet (formerly *Dryopithecus suevicus* Koken). Lower left molar from Trochtelfingen (FRG). c, d *Anthropodus brancoi* Schlosser 1901 *sive Neopithecus brancoi* Abel 1902 *sive Dryopithecus rhenanus* Remane 1921 *sive Dryopithecus brancoi,* possibly this fossil belongs to *Pliopithecus,* a hypothesis refuted by Kretzoi (1976). Site: Salmendingen, Swabia (FRG). e, f *Rudapithecus hungaricus* Kretzoi, Ind No. 46 from Rudabanya, Hungary

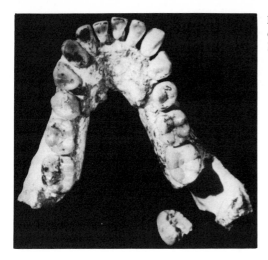

Fig. 3. *Ouranopithecus macedoniensis*
de Bonis et Melentis 1976. Mandible from
Bathylakkos Ind No. RPL 55, occlusal view

Fig. 4. *Ouranopithecus macedoniensis;*
enamel prism pattern of the molar in
a magnification of about 3000 x
(relating to the 10 x 10 scale)

Acknowledgments. This study was supported by a grant from the Deutsche Forschungsgemein-
schaft. We are also grateful to Prof. I. Melentis, Prof. M. Kretzoi and Dr. A. Czarnetzky for their help.

References

Andrews P (1978) A revision of the Miocene hominoidea of East Africa. Bull Br Mus Nat Hist
    (Geol) 30 (2):85–224
Bonis L de, Melentis J (1976) Les Dryopithecinés de Macédonie (Grèce): leur place dans l'évolu-
    tion des primates hominoides du Miocène. UISPP IXe Congr Colloq VI, Nice. (dir.: Tobias PV,
    Coppens Y), pp 26–38

Bonis L de, Bouvrain G, Geraads D, Melentis J (1974) Première découverte d'un primate homi-noide dans la Miocène supérieur de Macédonie (Grèce). CR Acad Sci Paris 278 D:3063–3066

Boyde A (1965a) The structure of developing mammalian dental enamel. In: Stack MV, Fearn-head RW (eds) Tooth enamel. J Wright and Sons, Bristal, pp 163–167

Boyde A (1965b) The development of enamel structure in mammals. In: Fleisch H, Blackwood H, Ower M (eds) Calcified tissues. Springer, Berlin Heidelberg New York, pp 276–280

Boyde A (1969) Electron microscopic observations relating to the nature and development of prism decussation in mammalian dental enamel. Bull Group Int Rech Sci Stomatol 12:151–208

Boyde A (1971) Comparative histology of mammalian teeth. In: Dahlberg AA (ed) Dental mor-phology and evolution. Univ Chicago Press, Chicago, pp 81–94

Carter JT (1922) The structure of the enamel in the primates and some other mammals. Proc Zool Soc London, pp 599–608

Gantt DG, Cring D (1979) Hominoid enamel prism patterns. Am J Phys Anthropol 50:440

Gantt DG, Pilbeam D, Stewart GP (1977) Hominoid enamel prism patterns. Science 198:1155–1157

Greenfield LO (1980) A late divergence hypothesis. Am J Phys Anthropol 52:351–365

Gustafson G, Gustafon A-G (1967) Microanatomy and histochemistry of enamel. In: Miles AEW (ed) Structural and chemical organisation of teeth. Academic Press, London New York, pp 75–134

Helmcke J-G (1967) Ultrastructure of enamel. In: Miles AEW (ed) Structural and chemical organi-sation of teeth. Academic Press, London New York, pp 135–163

Henke W (1981) Zum Ursprung der Hominidae. Naturwissenschaften

Henke W, Rothe H (1980) Der Ursprung des Menschen. Unser gegenwärtiger Wissensstand. G Fischer, Stuttgart

Kretzoi M (1975) New Ramapithecines and *Pliopithecus* from the Lower Pliocene of Rudabánya in North-Eastern Hungary. Nature (London) 257:578–581

Kretzoi M (1976) Die Hominisation und die Australopithecinen (in hung.). Anthropol Közl 20:3–11

Olshan AF (1979) A taxonomic study of *Dryopithecus macedoniensis*. Am J Phys Anthropol 50:468

Osborn JW (1970) The mechanism of prism formation in teeth: A hypothesis. Calcif Tissue Res 5:115–132

Pilbeam DR, Meyer GE, Badgley C, Rose MA, Pickford MHL, Behrensmeyer AK, Ibrahim Shah SM (1977) New hominoid primates from the Siwaliks of Pakistan and their bearing on homi-noid evolution. Nature (London) 270:689–697

Protsch RRR (1981) Monograph Garusi – The Garusi hominids I, II, and III. A new morpho-logical analysis based on a reconstruction, description, and dating of the Garusi hominids. In: Müller-Beck H (ed) Die archäologischen und anthropologischen Ergebnisse der Kohl-Larsen Expeditionen nach Nord-Tanzania 1933–1939. Archaeologica Venotoria, Tübingen

Regan CT (1930) Cited in Gantt et al (1977)

Shobusawa M (1952) Vergleichende Untersuchung über die Form der Schmelzprismen der Säuge-tiere. Okajimas Folia Anat Jpn 24:371–392

Simons EL (1961) The phyletic position of *Ramapithecus*. Postilla 57:1–9

Simons EL (1972) Primate evolution. An introduction to man's place in nature. Macmillan Comp, New York

Simons EW (1977) *Ramapithecus*. Sci Am 236:28–35

Simons EW, Pilbeam D (1965) Preliminary revision of the Dryopithecinae (Pongidae, Anthropoi-dea). Folia Primatol 3:81–152

Tomes J (1848) Lectures on dental physiology and surgery. Parker and Son, London

Vbra ES, Grine FE (1978) Australopithecine enamel prism patterns. Science 202:890–892

Xirotiris NI, Henke W, Symeonidis N (1979) Der M_3 von Megalopolis – ein Beitrag zu seiner morphologischen Kennzeichnung. Z Morphol Anthropol 70:117–122

The Structural Organization of the Cortex of the Motor Speech Areas of the Human Brain and Homologs on the Ape's Brain

M.S. Vojno[1]

Cyto- and myeloarchitectonics of the lower frontal convolution of the human brain (44 and 45 areas of Brodman) (25 hemispheres) have been studied plus homologous structures in the cortex of the brain of the higher primates — chimpanzee (14 hemispheres) and orangutan (12 hemispheres). We have prepared a parallel series of histological preparations (according to Nissle and Spilmeyer) in gelatinous modification. Each series includes 10 to 30–40 preparations. We measured the following: the width of the cortex and of each stratum; longitudinal and latitudinal sizes of the pyramidal cells in the stratums and substratums III', III'', III''', V', V'', and VI in both areas of each specimen; the number of pyramidal cells on such levels per unit of volume of the cortex; the number of radial myelinized clusters and fibers, their thickness (caliber) in the substratums III', V', V'', and in stratum VI. The average number of measurements in each object was from about 200–500 to 2,000–3,000, or even up to 10,000–20,000. Methods of measurement and biometrical modes of analysis have been published before (Vojno 1978).

At present there is no question that there are cortical structures in the brain of higher primates, which are homologous to the structures of the motor speech areas (44 and 45) of the human brain (Kononova 1962, Vojno 1972, 1978). It is shown by a whole complex of signs, such as: the average measurements of body cells; the pyramidal cells are arranged in correct vertical columns; the largest cells of substratum III''' are united into nest-like groupings; the proportions between the longitudinal and latitudinal sizes of the pyramidal cells and the proportions between the sizes of the cells in the same levels are very similar in all three species; the proportions of the width of the different cortex stratums are also similar. The absolute sizes of the pyramidal cells of the human brain and the brain of the chimpanzee are very similar: they cross each other.

The greatest differentiation in the cortex structure of the human and chimpanzee brains consists in the number of cells in a unit of volume in all the stratums and in both areas. In the chimpanzee brain cell density is 1.5 times greater than in the human brain. The brain of the orangutan has smaller cells and greater density than the chimpanzee brain.

The study of the human brain and the brain of the anthropomorphs shows us that there are two lines in their comparison. The first line demonstrates similar features,

1 Chair of Anthropology, Biology Department, Moscow University, Moscow, USSR

that is, the basis of the homologous complex. On the other hand, the lower frontal convolution of the human brain provides the unique function of Man's speech, which is absent in the apes. This functional uniqueness could not help being reflected in the morphological structure. There are many distinguishing features between these specimens at the macroanatomical level which we have noted earlier (Vojno 1978).

At the cytoarchitectonic level the main distinction between them (Man and ape) consists of two principal features: the density of cells and the surface area of the cortex of the homologous cytoarchitectonic structures. The last feature is very different in both specimens (Kononova 1962). If we assume a part of the cortex as a cube the side of which will be equal to the width of the cortex, we see that its size, the size of the cells in the cube, their topographical position and their proportions will be very similar in all the specimens. But the numbers of cubes in the human brain are more numerous than in the ape's brain. The density of cells in the human brain, on the contrary, is smaller. The decrease of neuron density in a unit of volume in the human brain as compared with the ape's brain shows that the first species has more connections than the other. They form rich plexi between cell bodies, which contain branching of dendrites and axons with their ramifications.

The brain of the orangutan has a specific structure of the cortex. It has the smallest surface area, smaller cortex width, smaller cell size, and the greatest density.

Myeloarchitectonics demonstrate both the similar as well as the different features in the cortex organization in all three species. In all three species the radial system belongs to the euradial type (by Vogt); the common number of the radial myelinized fibers in a unit of volume on the same level are similar too. On the other hand, the fibrillar structure of Man is much coarser than the fibrillar structure of the ape brain. The range of calibers of the radial myelinized fibers in the human brain ranges from 0.5 to 2.5 (microns) on the stratum III$'''$, and from 0.5 to even 4.5–5.0 on the lower stratums, V$''$ and VI. The last thickness is not found in the ape's brain. Clusters are wide in the human brain; there are many fibers in each one of them with different thicknesses (calibers), but the number of clusters in a unit of volume is smaller than in the ape's brain. If we represent one fiber as a cable and a cluster as the main line which connects many cables, then we seen in the human brain few wide lines with many different cables in each of them, including very rough ones. The ape's brain has many thin and compact lines with similar delicate cables in each of them. The spectrum of calibers is narrow and poor. The ape's structure is monotonous. But the number of clusters in the ape's brain in a unit of volume is, on the contrary, greater than in the human brain. Now we see two types of fibrillar organizations, and evidently Man's organization is higher than the ape's fibrillar structure. The nonradial fibrillar system belongs to the bistratum variants. But the plexi of the high levels, which belong to the associative complex and connect in the human brain with the higher specific human functions, are not as rich in the ape's brain.

Earlier we have expressed the idea of the hypothetical cortex structure which could be the basis for two nervous organizations, pongid and hominid, before their divergence into two different branches of evolutionary development (Vojno 1978).

It is not accidental that especially the lower frontal convolution has turned into a region which has provided the function of motor speech. The above is due to its topographical location on the brain. Its localization lies at the crossing of three morphological and functional systems within the entire frontal lobe. Behind it we find motor areas (four and six). Below it there is the basic cortex, regulating the activity of behavior and sleep, and which is responsible for emotions. From above and in front of this region there are prefrontal structures which carry responsibility for the higher forms of human behavior. The lower frontal convolution has a double duty. It works as a motor cortex but in its higher sphere: articulation. At the same time it codes in the sound the vocal signal, which contains the meaning or idea. Therefore, this region belongs to the prefrontal complex, but not to the motor cortex. Only the lower frontal convolution was able to accomplish, of itself, such unique higher function as human speech.

References

Komonova EP (1962) The frontal area of the brain. Medgiz Publishers, Moscow

Vojno MS (1972) On the homology of the cortical structures of the lower frontal convolution in the brain of man and the chimpanzee: "Man" (a collection of articles on the evolution and intra-specific differentiation). Mauka Publ, Moscow

Vojno MS (1978) The structure and development of the cortex of the lower frontal convolution of the human brain. X International Congress of Anthropological and Ethnological Sciences. Delhi, India, December 10–21, 1978

Behavioral Ecology and Sociobiology

ISSN 0340-5443   Title No. 265

Managing Editor: H. Markl, Konstanz

Editors: B. Hölldobler, Cambridge, MA; H. Kummer, Zürich; J. Maynard Smith, Brighton; E. O. Wilson, Cambridge, MA

Advisory Editors: G. W. Barlow, Berkeley, CA; J. Brown, Albany, NY; E. L. Charnov, Salt Lake City, UT; J. H. Crook, Bristol; J. F. Eisenberg, Washington, DC; T. Eisner, Ithaca, NY; S. T. Emlen, Ithaca, NY; V. Geist, Calgary, Alberta; D. R. Griffin, New York, NY; W. D. Hamilton, Ann Arbor, MI; D. von Holst, Bayreuth; K. Immelmann, Bielefeld; W. E. Kerr, Ribeirão Preto, SP; J. R. Krebs, Oxford; M. Lindauer, Würzburg; P. Marler, New York, NY; G. H. Orians, Seattle, WA; Y. Sugiyama, Inuyama City, Aichi; R. L. Trivers, Santa Cruz, CA; C. Vogel, Göttingen; C. Walcott, Ithaca, NY

The electric eel *(Electrophorus electricus),* often reaching a size of 8 feet and a weight of 200 pounds, electrocutes its prey with a burst of high-voltage electricity, about 500 Hz. In contrast, the eel locates prey using an electric organ emitting low-voltage impulses. An important question confronting behavioralists and physiologists alike is whether these two electrical systems are interrelated in some fundamental way. An answer will not only reveal important information on the physiology of *E. electricus,* but will also further our understanding of the basic relationship that exists between the environment and neuron function.

In response to this and other ecological and sociobiological questions Springer-Verlag initiated the publication of **Behavioral Ecology and Sociobiology** in 1976. Since then the journal has become a major forum, publishing original research on the functions, mechanics, and evolution of ecological adaptations and emphasizing social behavior. An international board of editors and advisors, aided by numerous reviewers, guarantees the very highest standards. Contributions are welcomed from scientists the world over; publication is almost exclusively in English. Topics treated cover a broad range, including orientation in space and time, communication and all other forms of social behavior, behavioral mechanisms of competition and resource partitioning, predatory and antipredatory behavior, and theoretical analyses of behavioral evolution. Quantitative studies carried out on representatives of nearly all major groups of the animal kingdom are published, from spiders through fish, birds, primates, and other mammals. Empirical studies on the biological basis of human behavioral adaptations are also welcomed.

Behavioral Ecology and Sociobiology has become an important source of information on the progress in animal behavior research. Scientists, researchers, and graduate students will regularly find papers of significance in their particular fields of interest.

For subscription information or sample copy write to:
Springer-Verlag, Journal Promotion Department,
P. O. Box 105280, D-6900 Heidelberg, FRG

Springer-Verlag
Berlin
Heidelberg
New York